It's another great book from CGP...

There are only three ways to make sure you're fully prepared for the new Grade 9-1 GCSE Biology exams — practice, practice and practice.

That's why we've packed this brilliant CGP book with realistic exam-style questions for every topic, and we've got all the practicals covered too.

And since you'll be tested on a wide range of topics in the real exams, we've also included a section of mixed questions to keep you on your toes!

CGP — still the best! ☺

Our sole aim here at CGP is to produce the highest quality books — carefully written, immaculately presented and dangerously close to being funny.

Then we work our socks off to get them out to you — at the cheapest possible prices.

Contents

Use the tick boxes to check off the topics you've completed.

Topic 7 — Ecology

Mixed Questions

Published by CGP

Editors:
Charlotte Burrows, Katherine Faudemer, Chris McGarry, Sarah Pattison.

Contributors:
Sophie Anderson, Helen Brace, James Foster, Bethan Parry, Alison Popperwell.

With thanks to Hayley Thompson and Phil Armstrong for the proofreading.

www.cgpbooks.co.uk
Clipart from Corel®
Printed by Elanders Ltd, Newcastle upon Tyne

Based on the classic CGP style created by Richard Parsons.

How to Use This Book

- Hold the book upright, approximately 50 cm from your face, ensuring that the text looks like this, not this. Alternatively, place the book on a horizontal surface (e.g. a table or desk) and sit adjacent to the book, at a distance which doesn't make the text too small to read.
- In case of emergency, press the two halves of the book together firmly in order to close.
- Before attempting to use this book, familiarise yourself with the following safety information:

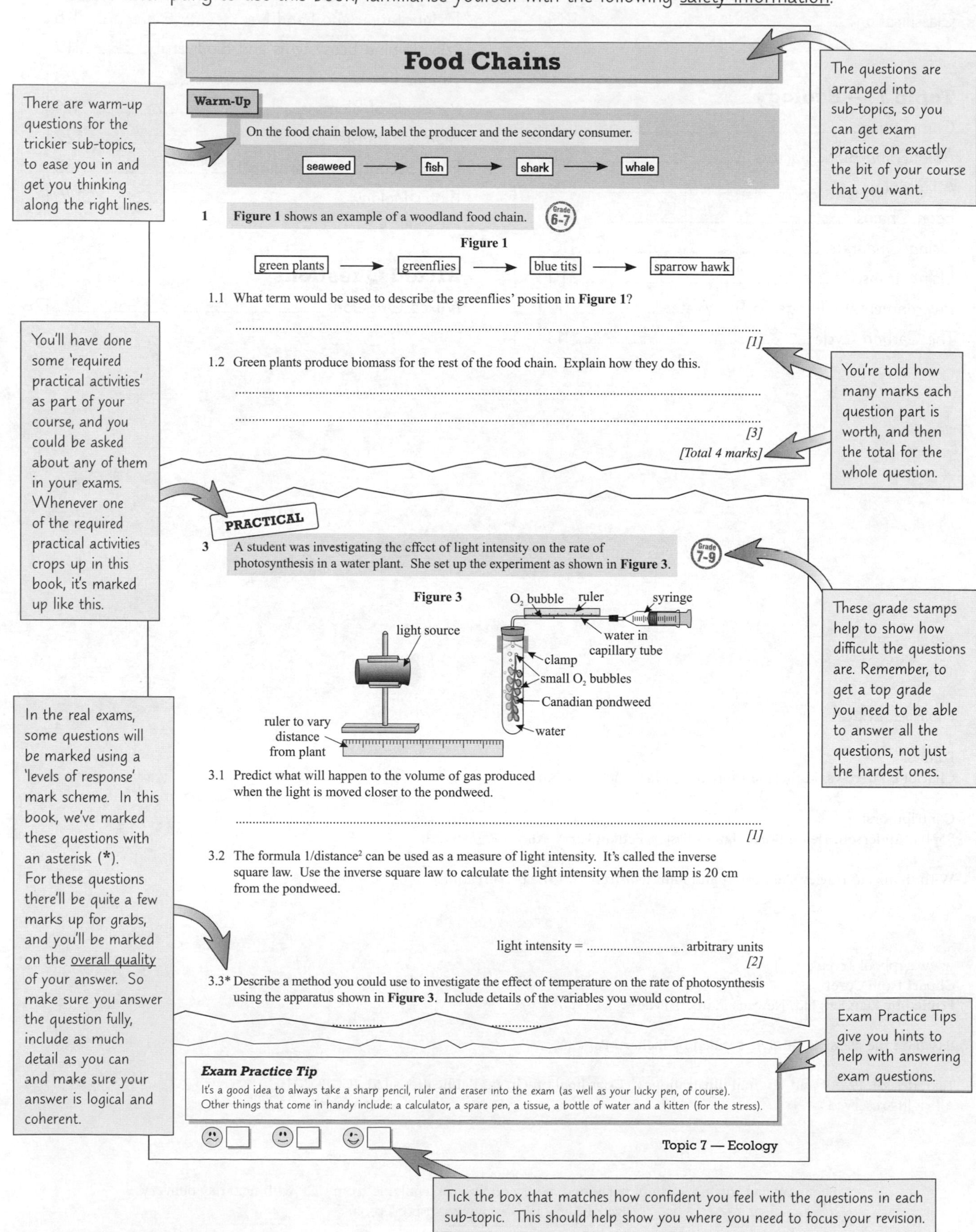

Food Chains
Warm-Up
On the food chain below, label the producer and the secondary consumer.
seaweed → fish → shark → whale
1 Figure 1 shows an example of a woodland food chain.
Grade 6-7
Figure 1
green plants → greenflies → blue tits → sparrow hawk
1.1 What term would be used to describe the greenflies' position in Figure 1?
[1]
1.2 Green plants produce biomass for the rest of the food chain. Explain how they do this.
[3]
[Total 4 marks]
PRACTICAL
3 A student was investigating the effect of light intensity on the rate of photosynthesis in a water plant. She set up the experiment as shown in Figure 3.
Grade 7-9
Figure 3
O_2 bubble
ruler
syringe
light source
water in capillary tube
clamp
small O_2 bubbles
Canadian pondweed
ruler to vary distance from plant
water
3.1 Predict what will happen to the volume of gas produced when the light is moved closer to the pondweed.
[1]
3.2 The formula 1/distance² can be used as a measure of light intensity. It's called the inverse square law. Use the inverse square law to calculate the light intensity when the lamp is 20 cm from the pondweed.
light intensity = ... arbitrary units
[2]
3.3* Describe a method you could use to investigate the effect of temperature on the rate of photosynthesis using the apparatus shown in Figure 3. Include details of the variables you would control.
Exam Practice Tip
It's a good idea to always take a sharp pencil, ruler and eraser into the exam (as well as your lucky pen, of course). Other things that come in handy include: a calculator, a spare pen, a tissue, a bottle of water and a kitten (for the stress).
Topic 7 — Ecology
There are warm-up questions for the trickier sub-topics, to ease you in and get you thinking along the right lines.
The questions are arranged into sub-topics, so you can get exam practice on exactly the bit of your course that you want.
You'll have done some 'required practical activities' as part of your course, and you could be asked about any of them in your exams. Whenever one of the required practical activities crops up in this book, it's marked up like this.
You're told how many marks each question part is worth, and then the total for the whole question.
These grade stamps help to show how difficult the questions are. Remember, to get a top grade you need to be able to answer all the questions, not just the hardest ones.
In the real exams, some questions will be marked using a 'levels of response' mark scheme. In this book, we've marked these questions with an asterisk (*). For these questions there'll be quite a few marks up for grabs, and you'll be marked on the overall quality of your answer. So make sure you answer the question fully, include as much detail as you can and make sure your answer is logical and coherent.
Exam Practice Tips give you hints to help with answering exam questions.
Tick the box that matches how confident you feel with the questions in each sub-topic. This should help show you where you need to focus your revision.

Cells

Warm-Up

Use the words on the right to correctly fill in the gaps in the passage.
You don't have to use every word, but each word can only be used once.

Most eukaryotic organisms are made up of .... cells.
They include .... and .... cells.
Prokaryotic organisms are .... cells. They are
.... and .... than eukaryotic cells.

many
smaller
plant
bacterial
single
larger
animal
simpler

1 **Figure 1** shows a diagram of an animal cell. 

Figure 1

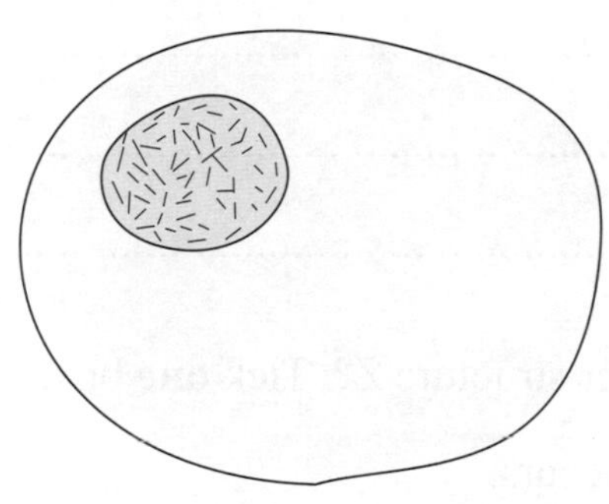

1.1 Label the cell membrane, cytoplasm and nucleus on **Figure 1**.

[3]

1.2 Give the function of each part of the cell on **Figure 1**.

Cell membrane ....

Cytoplasm ....

Nucleus ....

[3]

1.3 Name **two** other subcellular structures that can be found in an animal cell.
Describe the function of each structure.

....

....

....

....

[4]

1.4 Give **one** reason why the diagram in **Figure 1** does not represent a plant cell.

....

[1]

[Total 11 marks]

2 **Figure 2** shows a diagram of a prokaryotic cell.

Figure 2

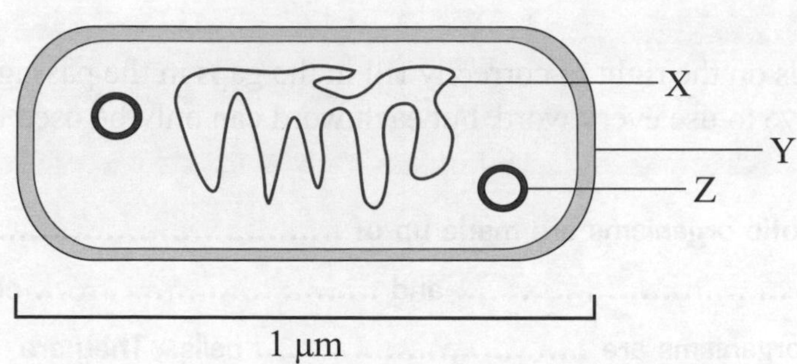

2.1 Which of the following is a prokaryotic cell?
Tick **one** box.

☐ root hair cell ☐ bacterium ☐ sperm ☐ nerve cell

[1]

2.2 Name structures X, Y and Z on **Figure 2**.

X ....

Y ....

Z ....

[3]

2.3 Which of the following is true for structure Z? Tick **one** box.

☐ It is where photosynthesis occurs.

☐ It is part of the cell membrane.

☐ It contains genetic material.

[1]

Look at the scale on **Figure 2**.

2.4 A eukaryotic cell measures 10 μm long.
How many times larger is it than the cell in **Figure 2**?

....

[1]

2.5 The head of a pin is approximately 1 mm in diameter.
How many prokaryotic cells would fit lengthways across it?

.... cells

[2]

2.6 Give **one** difference between prokaryotic and eukaryotic cells, other than their size.

....

....

[1]

[Total 9 marks]

Microscopy

1 A student observed blood cells under a microscope.
A scale drawing of one of the cells is shown in **Figure 1**.

Figure 1

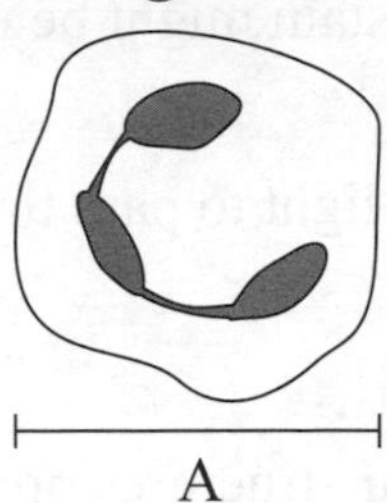

In **Figure 1**, A is the image width. The real width of the cell is 0.012 mm.
What is the magnification of the image? Use the formula:

$$\text{magnification} = \frac{\text{size of image}}{\text{size of real object}}$$

magnification = × ....

[Total 2 marks]

2 A plant cell is magnified 1000 times using a light microscope.

2.1 The length of the image of the plant cell is 10 mm.
Calculate the actual length of one plant cell in millimetres (mm).
Use the formula:

$$\text{magnification} = \frac{\text{size of image}}{\text{size of real object}}$$

.... mm

[2]

2.2 What is the length of one plant cell in micrometres (μm)?

.... μm

[1]

2.3 How do magnification and resolution compare between electron and light microscopes?

....

....

[2]

2.4 Explain how electron microscopy has increased understanding of subcellular structures.

....

....

[2]

[Total 7 marks]

PRACTICAL

More on Microscopy

1 A student wants to use a light microscope to view a sample of onion cells. Grade 4-6

1.1 The student adds a drop of iodine stain to her sample.
Which statement best describes when a stain might be used to view a sample of tissue?
Tick **one** box.

☐ When the specimen is too thick for light to pass through.

☐ When the specimen is colourless.

☐ When there aren't many sub-cellular structures present in the cells.

☐ When a cover slip is not being used.

[1]

Figure 1 shows a diagram of the light microscope that the student plans to use.

Figure 1

1.2 The three different objective lenses are labelled in **Figure 1** with their magnification. Which lens should the student select first when viewing her cells?

....

[1]

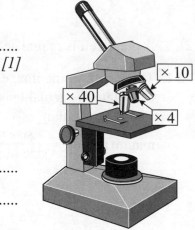

1.3 After she has selected the objective lens, she looks down the eyepiece and uses the adjustment knobs. Describe the purpose of the adjustment knobs.

....

....

....

[1]

1.4 The student wants to see the cells at a greater magnification.
Describe the steps that she should take.

....

....

....

[2]

1.5 After she has viewed the cells, she wants to produce a scientific drawing of them.
Her teacher has advised her to use clear, unbroken lines to draw the structures she can see.
Give **two** other ways in which she can ensure she produces an accurate and useful drawing.

1. ....

2. ....

[2]

[Total 7 marks]

Cell Differentiation and Specialisation

Warm-Up

Different types of cell have different structures that help them carry out specific functions.
Draw arrows below to match up each type of plant cell with its structure and function.

Plant cell	Structure and Function
root hair cell	Very few subcellular structures and holes in the end cell walls allow dissolved sugars to move from one cell to the next.
xylem	Lots of chloroplasts for absorption of sunlight.
phloem	Cells that are hollow in the centre and have no end cell walls form a continuous tube for transporting water from roots to leaves.
	Long finger-like projection increases surface area for absorption of water.

1 As an organism develops, some of its cells develop different structures and change into different types of cells. This allows the cells to carry out specific functions. What is this process called? Tick **one** box.

Grade 4-6

☐ mitosis ☐ adaptation ☐ differentiation ☐ specialisation

[Total 1 mark]

2 Sperm cells are specialised to help them achieve their function. **Figure 1** shows the structure of a sperm cell.

Grade 6-7

Figure 1

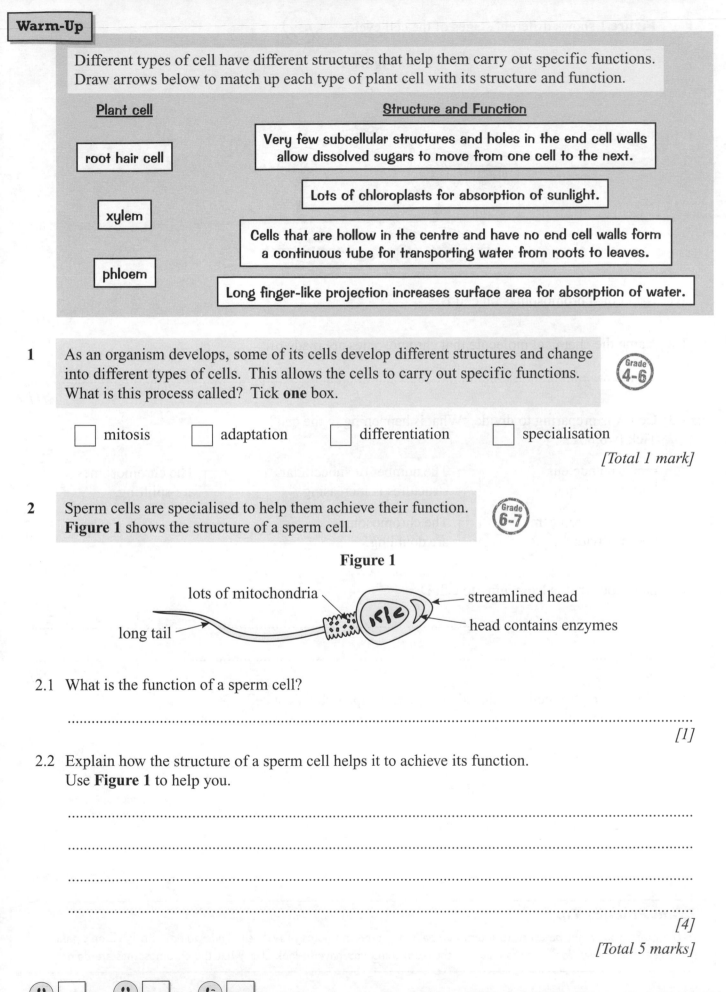

2.1 What is the function of a sperm cell?

....

[1]

2.2 Explain how the structure of a sperm cell helps it to achieve its function. Use **Figure 1** to help you.

....

....

....

....

[4]

[Total 5 marks]

Chromosomes and Mitosis

1 **Figure 1** shows different stages of the cell cycle. 

Figure 1

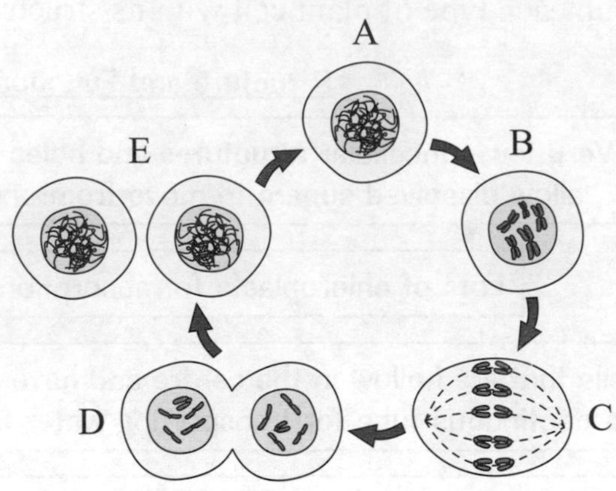

1.1 Label the chromosomes on cell B.

[1]

1.2 Name the chemical molecule that chromosomes are made of.

....

[1]

1.3 Cell A is preparing to divide. What is happening to the cell?
Tick **two** boxes.

☐ The nucleus is dividing.

☐ The number of subcellular structures is increasing.

☐ The chromosomes are splitting.

☐ The cytoplasm is dividing.

☐ The chromosomes are doubling.

[2]

1.4 Describe what is happening to cell D.

....

....

[2]

1.5 How do the two cells produced at stage E compare to parent cell A?
Tick **one** box.

☐ They are genetically different.

☐ They are genetically similar.

☐ They are genetically identical.

[1]

[Total 7 marks]

Exam Practice Tip

In the exam, you might be asked to interpret what's going on in photos of real cells undergoing mitosis. Don't panic if the cells themselves don't look familiar — the main thing you have to look at is what the chromosomes are doing.

Binary Fission

1 Scientists performed an experiment to observe the growth of bacteria in a nutrient broth at 25 °C. **Table 1** shows their results.

Table 1

Time (hours)	0	2	4	6	8	10	12	14	16
Absorbance (au)	0.01	0.05	0.10	0.20	0.40	0.66	0.78	0.80	0.80

1.1 Use the data in **Table 1** to finish plotting **Figure 1** and draw a line of best fit.

Figure 1

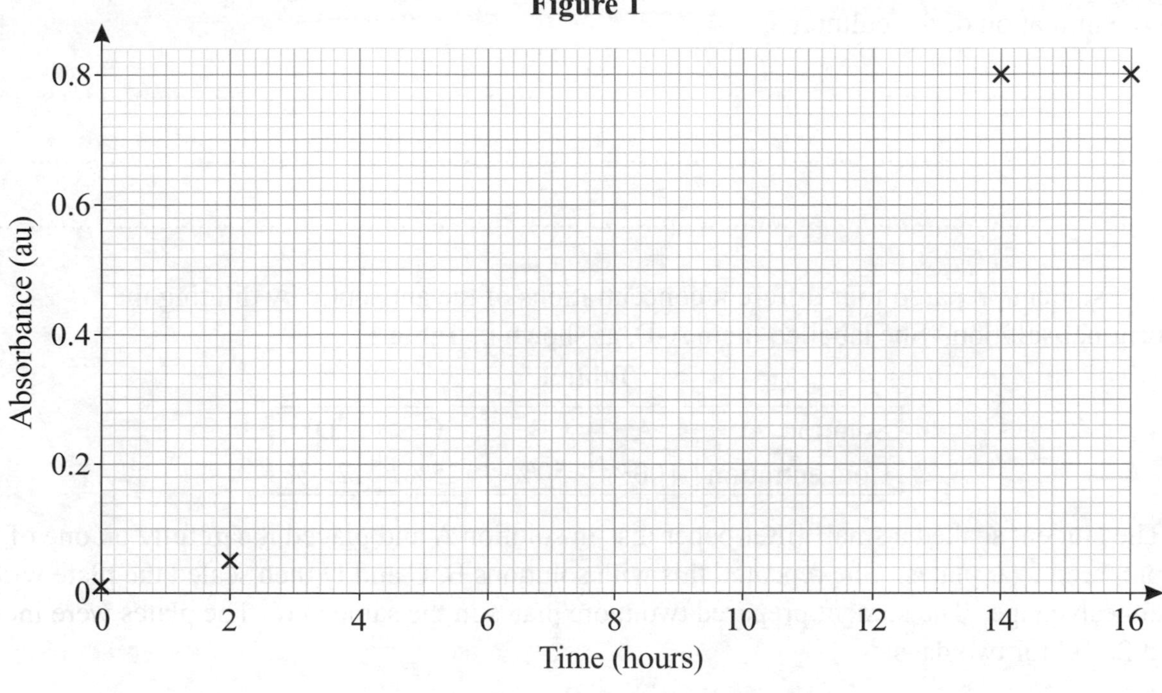

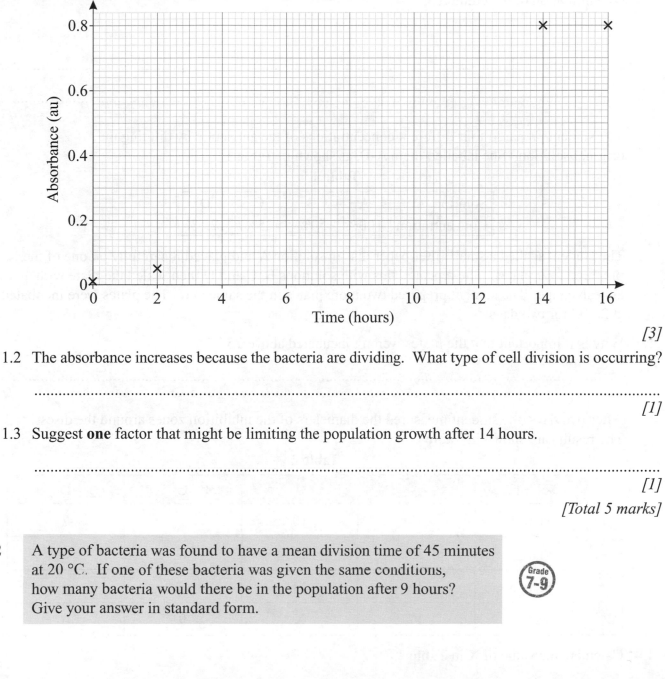

[3]

1.2 The absorbance increases because the bacteria are dividing. What type of cell division is occurring?

....

[1]

1.3 Suggest **one** factor that might be limiting the population growth after 14 hours.

....

[1]

[Total 5 marks]

2 A type of bacteria was found to have a mean division time of 45 minutes at 20 °C. If one of these bacteria was given the same conditions, how many bacteria would there be in the population after 9 hours? Give your answer in standard form.

Number of bacteria = ...

[Total 4 marks]

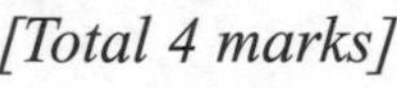

PRACTICAL Culturing Microorganisms

1 A student was investigating the effect of different concentrations of an antiseptic on the growth of *Bacillus subtilis* bacteria.

Grade 6-7

1.1 The student prepared uncontaminated cultures of the bacteria on agar plates. Why was it important that the cultures were not contaminated?

....

.... *[2]*

1.2 Suggest **three** things that the student may have done while preparing the plates to prevent contamination of the cultures.

....

....

.... *[3]*

The student prepared four different concentrations of the antiseptic by diluting the original solution. She labelled them A-D, as shown in **Table 1**.

Table 1

Solution	A	B	C	D
Concentration	100%	50%	25%	12.5%

The student soaked a small filter paper disc in solution A and placed it carefully on one of the prepared agar plates. She repeated this with solutions B, C and D then sealed the plate with adhesive tape. The student prepared two more plates in the same way. The plates were incubated at 25 ˚C for two days.

1.3 Why is it important that the plates weren't incubated above 25 ˚C?

.... *[1]*

After two days, the student measured the diameters of the inhibition zones around the discs. The results are shown in **Table 2**.

Table 2

	Solution	A	B	C	D
Diameter (mm)	Plate 1	17	10	7	7
	Plate 2	19	15	8	5
	Plate 3	15	14	12	6
	Mean	17	X	9	6

1.4 Calculate the value of **X** in **Table 2**.

X = ... mm *[1]*

The student repeated the experiment using a different antiseptic.
The results are summarised in **Table 3**.

Table 3

Concentration of solution	100%	50%	25%	12.5%
Mean radius (mm)	11	8	5	4
Area of inhibition zone (cm^2)	3.8	2.0	Y	0.5

1.5 Calculate the value of Y to complete **Table 3**.
Use the formula πr^2, where $\pi = 3.14$. Give your answer in cm^2.

Y = ... cm^2
[2]

1.6 Use **Table 3** to plot a graph of the area of the inhibition zone against the concentration of the solution for the 100%, 50% and 12.5% solutions. Draw a line of best fit between the points.

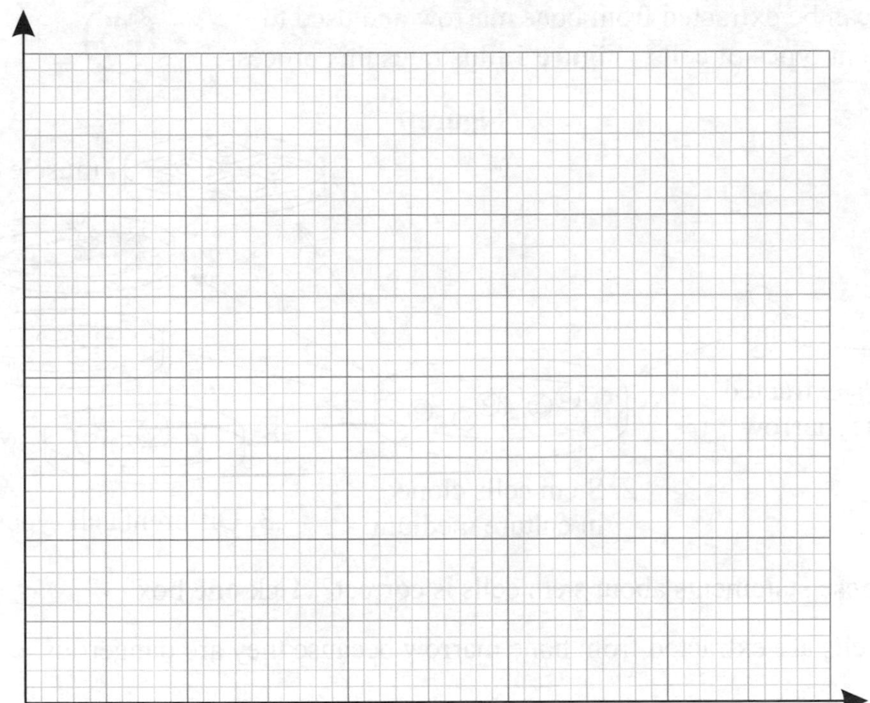

[3]

1.7 Use the graph to estimate the area of the inhibition zone produced by a solution of this antiseptic with a concentration of 80%.

....
[1]

1.8 Describe the relationship between the concentration of antiseptic and its effectiveness at preventing bacterial growth.

....
[1]

[Total 14 marks]

Stem Cells

1 Stem cells can be found in the growing areas of plants. Grade 4-6

1.1 What are these growing areas of a plant called?
Tick **one** box.

☐ cloning zones ☐ meristems ☐ leaves ☐ mesophyll layers

[1]

1.2 You can produce cloned plants from plant stem cells.
Describe **three** benefits of producing cloned plants from stem cells.

....

....

....

[3]

[Total 4 marks]

2 Stem cells can be extracted from bone marrow and used to grow different types of cells. **Figure 1** illustrates this process. Grade 6-7

Figure 1

1. Stem cells extracted from bone marrow.
2. Stem cells cloned in culture medium.
muscle cells
nerve cells
red blood cells
white blood cells
3. Different cell types are produced.

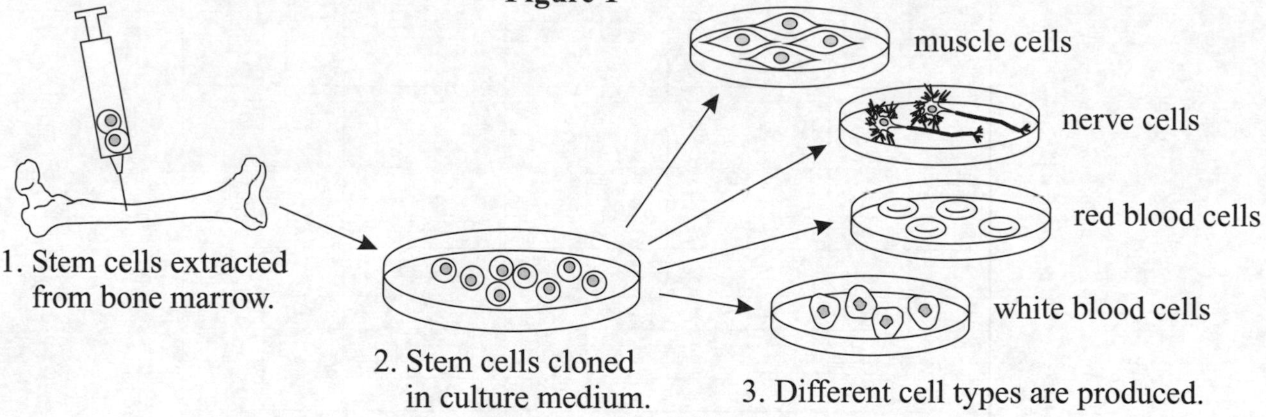

2.1 Which of these statements about stem cells is correct? Tick **one** box.

☐ Stem cells are extracted from bone marrow because they are dangerous.

☐ Stems cells are differentiated cells.

☐ Stem cells can be found in every organ of the body.

☐ Stem cells can differentiate into many types of body cell.

[1]

2.2 Why are the stem cells cloned?

....

[1]

2.3 Why can't all body cells be used to grow different types of cell?

....

....

[1]

The technique shown in **Figure 1** could be used to produce cells for some medical treatments.

2.4 Besides bone marrow, where else can stem cells for medical treatments be obtained from?

....

[1]

2.5 Name **one** medical condition that may be helped by treatment using stem cells.

....

[1]

2.6 Give **one** potential risk of using stem cells in medical treatments.

....

[1]

[Total 6 marks]

3 **Figure 2** shows the process of therapeutic cloning. Grade 7-9

Figure 2

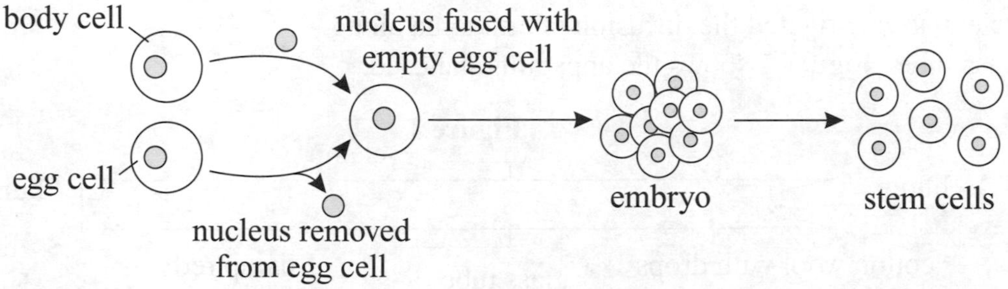

body cell

nucleus fused with empty egg cell

egg cell

nucleus removed from egg cell

embryo

stem cells

3.1 Describe what therapeutic cloning is.

....

[1]

3.2 Explain the benefit of using stem cells produced by therapeutic cloning for medical treatments compared to stem cells from a donor.

....

....

[2]

Therapeutic cloning involves creating an embryo, from which the stem cells for treatment are sourced. For this reason, some people are against using therapeutic cloning.

3.3* Discuss the ethical issues surrounding the use of embryonic stem cells in medicine and research.

....

....

....

....

....

[4]

[Total 7 marks]

Exam Practice Tip

If you're asked to write about social or ethical issues on a topic in your exam, it's a good idea to write down different points of view, so that you give a well-balanced answer. You don't have to agree with the opinions you write about.

Diffusion

Warm-Up

The diagram on the right shows three cells. The carbon dioxide concentration inside each cell is shown. Draw arrows between the cells to show in which directions the carbon dioxide will diffuse.

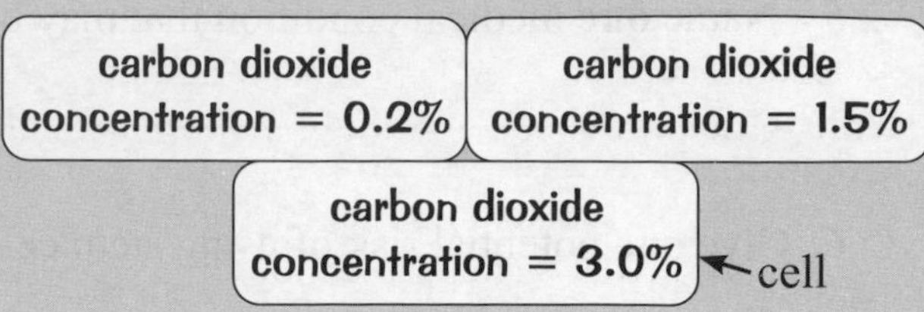

1 Which of these molecules is not able to diffuse through a cell membrane? Tick **one** box.

☐ protein ☐ oxygen ☐ glucose ☐ water

[Total 1 mark]

2 A scientist investigated the diffusion of ammonia along a glass tube. **Figure 1** shows the apparatus she used.

Figure 1

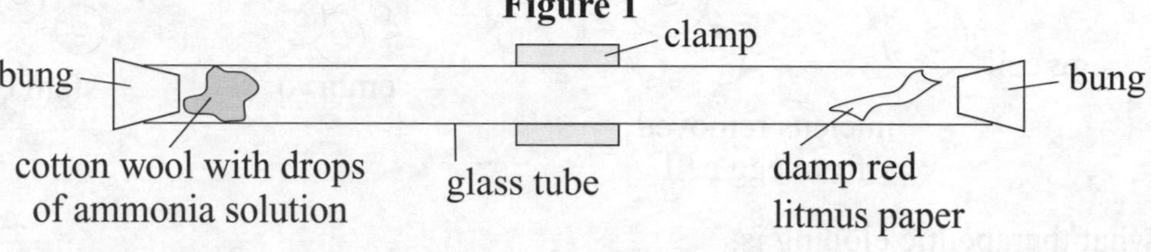

When the ammonia reaches the end of the tube, the litmus paper changes colour. The scientist timed how long this colour change took at five different concentrations of ammonia. **Table 1** shows her results.

Table 1

Concentration of ammonia (number of drops)	1	2	3	4	5
Time (s)	46	35	28	19	12

2.1 Define diffusion in terms of the particles of a gas.

....

....

[3]

2.2 What do the results in **Table 1** show about the effect of concentration on the rate of diffusion?

....

[1]

2.3 State **two** factors, other than concentration gradient, that affect the rate of diffusion into a cell.

....

....

[2]

2.4 Suggest how the scientist could increase the precision of her results.

....

[1]

[Total 7 marks]

Osmosis

1 Osmosis is a form of diffusion. Grade 4-6

1.1 Define osmosis.

....

....

.... *[3]*

1.2 In which of these is osmosis occurring? Tick **one** box.

- [ ] A plant is absorbing water from the soil.
- [ ] Sugar is being taken up into the blood from the gut.
- [ ] Water is evaporating from a leaf.
- [ ] Oxygen is entering the blood from the lungs.

[1]

[Total 4 marks]

PRACTICAL

2 A student did an experiment to see the effect of different salt solutions on pieces of potato. He cut five equal-sized rectangular chips from a raw potato and determined the mass of each chip. Each chip was placed in a beaker containing a different concentration of salt solution. The mass of each chip was measured again after 24 hours. The results are shown in **Table 1**. Grade 6-7

Table 1

Beaker	1	2	3	4	5
Concentration of salt solution (%)	0	1	2	5	10
Mass of potato chip at start of experiment (g)	5.70	5.73	5.71	5.75	5.77
Mass of potato chip after 24 hours (g)	6.71	6.58	6.27	5.46	4.63

2.1 Explain why it is important that all the potato pieces come from the same potato.

.... *[1]*

2.2 Calculate the percentage change in mass after 24 hours for the potato chip in beaker 2.

.... %

[2]

2.3 The student wanted to find a solution that would not cause the mass of the chip to change. Suggest what concentration of salt solution the student should try.

.... *[1]*

[Total 4 marks]

Active Transport

1 Sugar molecules can be absorbed from the gut into the blood by active transport. Grade 4-6

1.1 Define active transport.

....

....

[1]

1.2 State how sugar molecules are used inside cells.

....

[1]

1.3 Which of these statements about active transport is correct? Tick **one** box.

- [ ] It is the way in which oxygen enters the blood from the lungs.
- [ ] It can only occur down a concentration gradient.
- [ ] It needs energy from respiration.

[1]

[Total 3 marks]

2 Plants absorb mineral ions from the soil by active transport. Grade 6-7

2.1 Explain why plants need mineral ions.

....

[1]

2.2 Explain why plants are not able to rely on diffusion to absorb mineral ions from the soil.

....

....

[2]

2.3 State **two** ways in which active transport differs from diffusion.

....

....

[2]

2.4 Describe the function and structure of the root hair cells of a plant.
Include details of how the structure of the root hair cell helps it to carry out its function.

....

....

....

....

[3]

[Total 8 marks]

Exchange Surfaces

Warm-Up

Place the following organisms in order according to their surface area to volume ratio. Number the boxes 1 to 4, with 1 being the smallest and 4 being the largest.

☐ **Bacterium** ☐ **Tiger** ☐ **Domestic cat** ☐ **Blue whale**

1 Give **four** features of an effective gas exchange surface in an animal. Grade 4-6

....

....

....

....

[Total 4 marks]

2 A student was investigating the effect of size on the uptake of substances by diffusion. He cut different sized cubes of agar containing universal indicator and placed them in beakers of acid. The student timed how long it took for the acid to diffuse through to the centre of each cube (and so completely change the colour of the agar). Grade 7-9

Table 1 shows the relationship between the surface area and volume of the agar cubes.

Table 1

Cube size (cm)	Surface area (cm^2)	Volume (cm^3)	Simple ratio
$2 \times 2 \times 2$	24	8	3:1
$3 \times 3 \times 3$	**X**	**Y**	2:1
$5 \times 5 \times 5$	150	125	**Z** : 1

2.1 Calculate the values of X and Y in **Table 1**.

X = ... cm^2

Y = ... cm^3

[2]

2.2 Calculate the value of Z.

Z = ...

[1]

2.3 Predict which cube took the longest to change colour. Give **one** reason for your answer.

Cube ....

Reason ....

[1]

[Total 4 marks]

Exchanging Substances

1 **Figure 1** shows an alveolus in the lungs. Grade 4-6

Figure 1

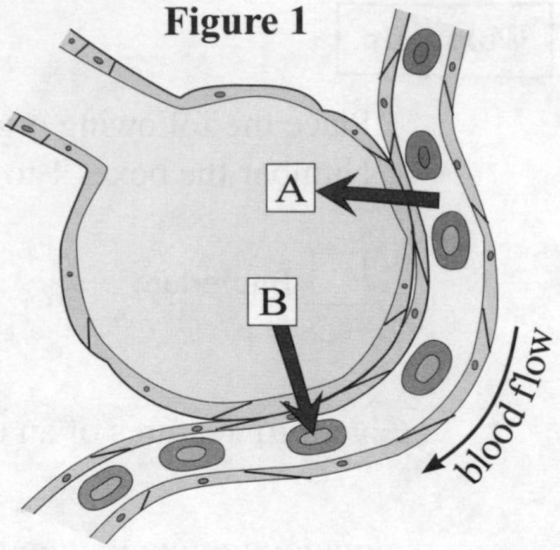

1.1 Name gases A and B.

A ....

B ....

[2]

1.2 By what process do these gases move across the membrane?

....

[1]

1.3 State which feature of the lungs gives them:

a short diffusion pathway ....

a large surface area ....

[2]

[Total 5 marks]

2 Emphysema is a disease that weakens and breaks down the walls of the alveoli. Grade 6-7

A person with emphysema may suffer from lower energy levels during physical exercise. Suggest and explain the cause of this symptom.

....

....

....

....

[Total 3 marks]

3 Describe and explain how the structure of the small intestine is adapted for absorbing the products of digestion. Grade 6-7

....

....

....

....

....

....

....

[Total 6 marks]

Exam Practice Tip

It may seem obvious, but if you're asked to explain how the structure of something relates to its function, don't just dive straight in and rattle off what it looks like. First, focus on the function being asked about, then pick out the individual structures that help to do that function and for each structure, make sure you give a clear explanation of how it helps.

More on Exchanging Substances

1 Leaves are adapted for gas exchange. **Figure 1** shows the cross-section of a leaf. Grade 4-6

1.1 Name the channels labelled X.

Figure 1

....

[1]

1.2 Describe the movement of gases into and out of the leaf.

....

....

....

[3]

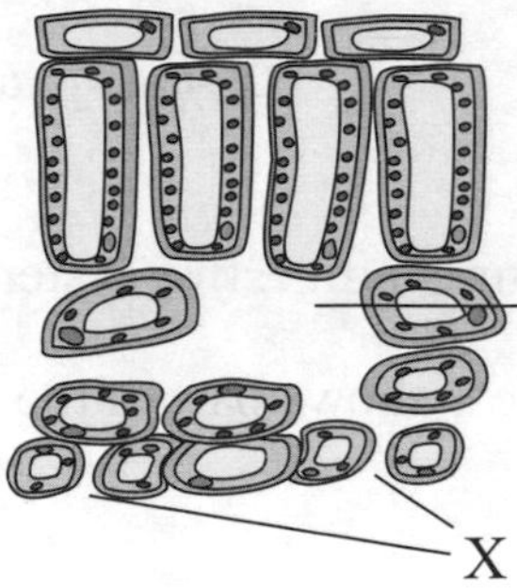

1.3 Suggest the purpose of the air spaces in the leaf.

....

....

[1]

[Total 5 marks]

2 **Figure 2** shows a diagram of a fish gill, which is a gas exchange surface. Grade 6-7

2.1 How do gill filaments increase the efficiency of the gas exchange surface?

Figure 2

....

....

[1]

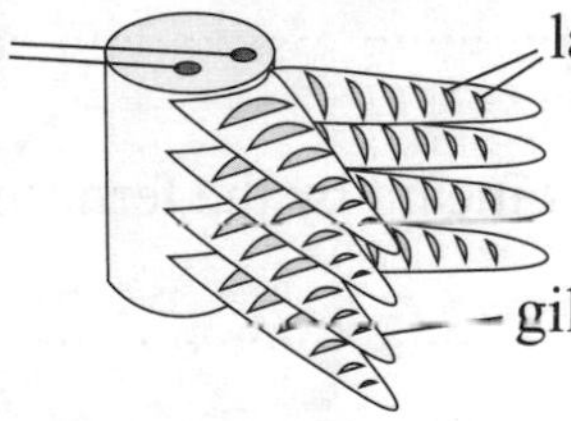

2.2 What is the purpose of the lamellae?

....

[1]

2.3 Describe one other feature of an efficient gas exchange surface that is present in **Figure 2**.

....

[1]

The number and length of gill filaments differ between types of fish.

2.4 Describe the differences in the gill filaments you would expect between a fast-moving fish and a slow-moving fish.

....

[1]

2.5 Explain why you would expect to see these differences.

....

....

[2]

[Total 6 marks]

Cell Organisation

Warm-Up

Number the boxes 1 to 4 to put the body components in order of size from smallest to largest, 1 being the smallest and 4 being the largest.

☐ **Organ system** ☐ **Tissue** ☐ **Cell** ☐ **Organ**

1 The human digestive system is an example of an organ system. Grade 4-6

1.1 **Figure 1** shows part of the digestive system.

Figure 1

Name organs X, Y and Z.

X: ....

Y: ....

Z: ....

[3]

1.2 What is meant by the term 'organ system'?

....

....

[1]

1.3 Organ systems contain multiple types of tissue.
What is a tissue?

....

....

[1]

1.4 What is the role of the digestive system?

....

[1]

1.5 The stomach is an organ that is part of the digestive system.
What is an organ?

....

....

[1]

[Total 7 marks]

Enzymes

1 Enzymes are biological catalysts. They increase the rate of biological reactions. **Figure 1** shows a typical enzyme.

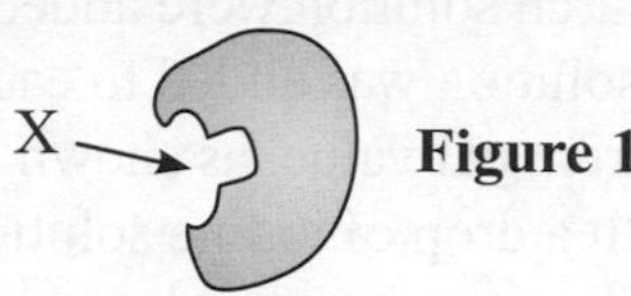

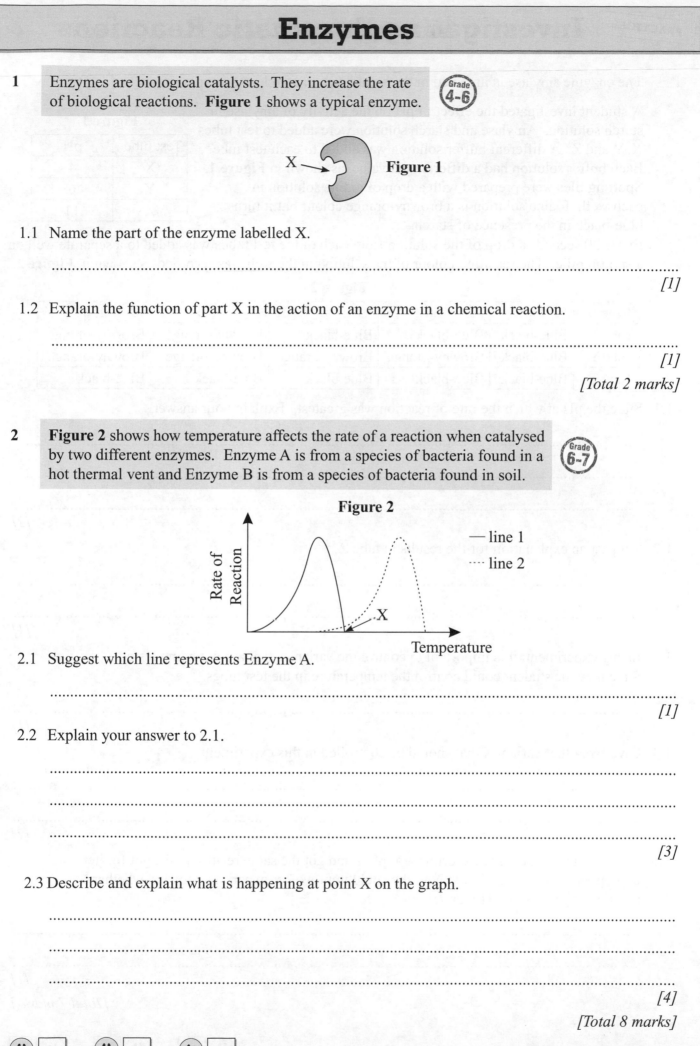

Figure 1

1.1 Name the part of the enzyme labelled X.

....

[1]

1.2 Explain the function of part X in the action of an enzyme in a chemical reaction.

....

[1]

[Total 2 marks]

2 **Figure 2** shows how temperature affects the rate of a reaction when catalysed by two different enzymes. Enzyme A is from a species of bacteria found in a hot thermal vent and Enzyme B is from a species of bacteria found in soil.

Grade 6-7

Figure 2

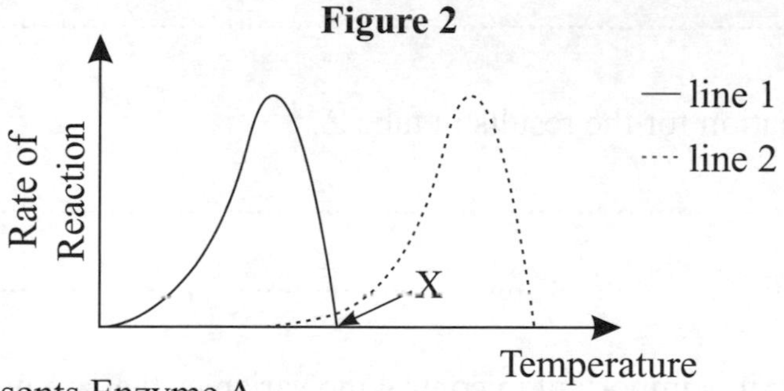

Rate of Reaction

Temperature

— line 1

···· line 2

X

2.1 Suggest which line represents Enzyme A.

....

[1]

2.2 Explain your answer to 2.1.

....

....

....

[3]

2.3 Describe and explain what is happening at point X on the graph.

....

....

....

[4]

[Total 8 marks]

PRACTICAL Investigating Enzymatic Reactions

1 The enzyme amylase is involved in the breakdown of starch into simple sugars.

A student investigated the effect of pH on the activity of amylase in starch solution. Amylase and starch solution were added to test tubes X, Y and Z. A different buffer solution was added to each test tube. Each buffer solution had a different pH value, as shown in **Figure 1**. Spotting tiles were prepared with a drop of iodine solution in each well. Iodine solution is a browny-orange colour but it turns blue-black in the presence of starch.

Figure 1

Test tube	pH
X	4
Y	6
Z	11

Every 30 seconds a drop of the solution from each of the test tubes was added to a separate well on a spotting tile. The resulting colour of the solution in the well was recorded as shown in **Figure 2**.

Figure 2

Time (s)	30	60	90	120	150
Tube X	Blue-black	Blue-black	Blue-black	Browny-orange	Browny-orange
Tube Y	Blue-black	Browny-orange	Browny-orange	Browny-orange	Browny-orange
Tube Z	Blue-black	Blue-black	Blue-black	Blue-black	Blue-black

1.1 State the pH at which the rate of reaction was greatest. Explain your answer.

....

....

....

[2]

1.2 Suggest an explanation for the results in tube **Z**.

....

....

[1]

1.3 In any experiment, it is important to control the variables that are not being tested. State how the student could control the temperature in the test tubes.

....

[1]

1.4 Give **two** other variables that should be controlled in this experiment.

1. ....

2. ....

[2]

1.5 The student repeated her experiment at pH 7 and got the same results as she got for her experiment at pH 6. Describe how she could improve her experiment to find whether the reaction is greatest at pH 6 or 7.

....

....

[1]

[Total 7 marks]

Enzymes and Digestion

Warm-Up

The diagram on the right shows some of the organs in the digestive system. Lipases and proteases are examples of digestive enzymes.

Write an 'L' on the organs that produce lipases and write a 'P' on the organs that produce proteases.

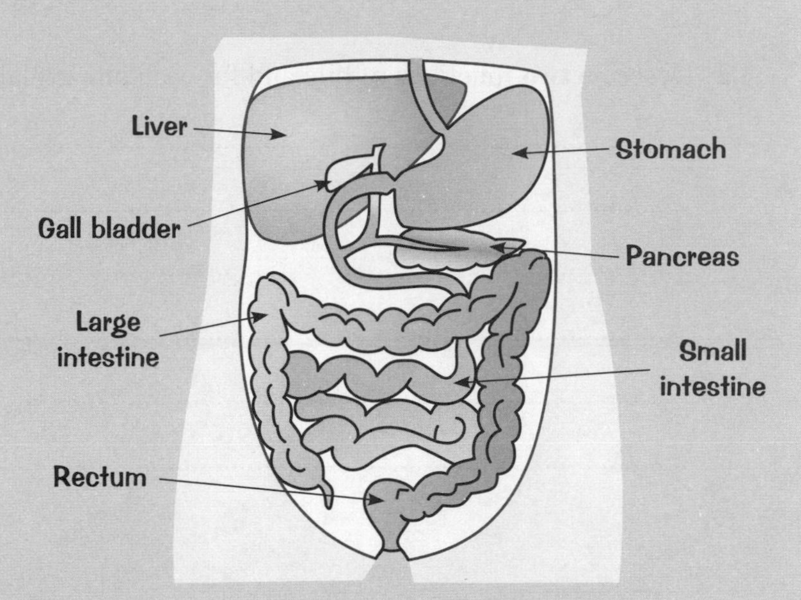

1 Amylase is a digestive enzyme. 

1.1 Which group of digestive enzymes does amylase belong to?
Tick **one** box.

☐ Carbohydrases ☐ Lipases ☐ Proteases

[1]

1.2 What is the product of the reaction catalysed by amylase?
Tick **one** box.

☐ Sugars ☐ Amino acids ☐ Glycerol ☐ Fatty acids

[1]

[Total 2 marks]

2 The process of digestion relies on the action of many different types of digestive enzyme. Grade 4-6

2.1 Describe the role of digestive enzymes in the process of digestion.

....

....

....

[2]

2.2 Give **two** ways in which the products of digestion can be used by the body.

....

....

[2]

[Total 4 marks]

3 Bile plays an important role in the digestive system. Grade 6-7

3.1 Name the organ where bile is produced and the organ where it is stored.

Produced .... Stored ....

[2]

3.2 Describe **two** functions of bile and for each one explain why it is important.

....

....

....

....

....

....

[4]

[Total 6 marks]

4* Different types of food molecule are broken down by different digestive enzymes. Using your knowledge of digestive enzymes and where they are produced in the body, fully outline the processes involved in the digestion of a meal containing carbohydrates, proteins and lipids. Grade 7-9

....

....

....

....

....

....

....

....

....

....

....

....

....

....

....

....

[Total 6 marks]

Food Tests

PRACTICAL

Warm-Up

Draw lines to connect the tests on the left with the biological molecules that they identify.

Biuret test | Benedict's test | Lipids | Proteins

Sudan III test | Iodine test | Starch

1* A student is analysing the nutrient content of egg whites.

Fully describe an investigation that the student could carry out to find out if protein is present in a sample of the egg whites.

....

....

....

....

....

....

....

....

[Total 6 marks]

2 A student was given test tubes containing the following glucose concentrations: 0 M, 0.02 M, 0.1 M, 1 M. The test tubes were not labelled and he was asked to perform tests to determine which test tube contained which glucose solution.

Grade 6-7

2.1 Describe the test he could carry out to try and distinguish between the glucose solutions.

....

....

....

[3]

2.2 **Table 1** shows the substance observed in the test tubes following his tests. Complete the table to show which glucose solution (0 M, 0.02 M, 0.1 M, 1 M) each test tube contained.

Table 1

	Tube 1	Tube 2	Tube 3	Tube 4
substance observed	yellow precipitate	blue solution	red precipitate	green precipitate
glucose concentration (M)	....	....	....	....

[1]

[Total 4 marks]

The Lungs

Warm-Up

Use the words on the right to correctly fill in the gaps in the passage.
You don't have to use every word, but each word can only be used once.

When you breathe in, air flows through a series of tubes, including the trachea and the .... The .... are where gas exchange takes place. Gas exchange .... the blood and allows the removal of .... from the body.

oxygenates
arteries
alveoli
carbon dioxide
bronchi
oxygen
deoxygenates

1 **Figure 1** shows the human respiratory system. Grade 4-6

Figure 1

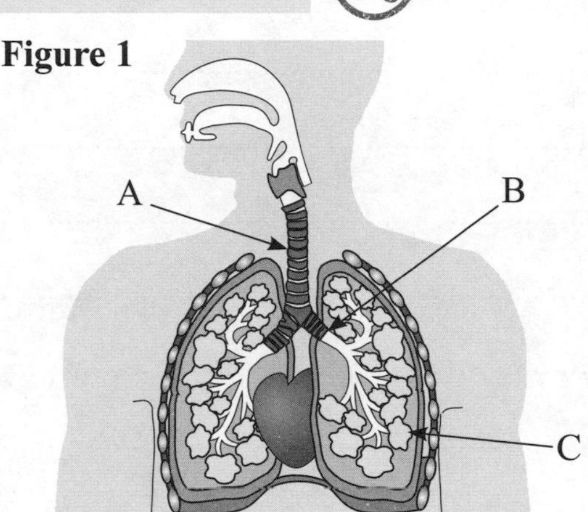

1.1 Name the parts labelled A, B and C in **Figure 1**.

A .... B .... C ....

[3]

Figure 2 shows a close-up of part C from **Figure 1**.

Figure 2

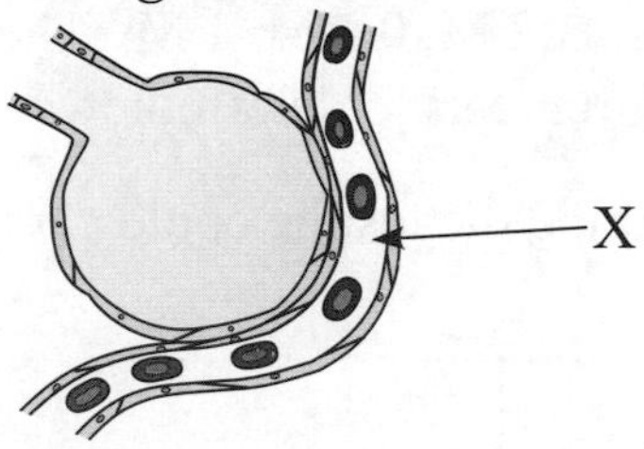

X

1.2 Name the structure labelled X in **Figure 2**.

....

[1]

1.3 Describe the role that structure X plays in gas exchange in the lungs.

....

....

....

....

[4]

[Total 8 marks]

Circulatory System — The Heart

1 Humans have a double circulatory system. The heart pumps blood around the body through a network of veins and arteries. **Figure 1** shows a diagram of the heart.

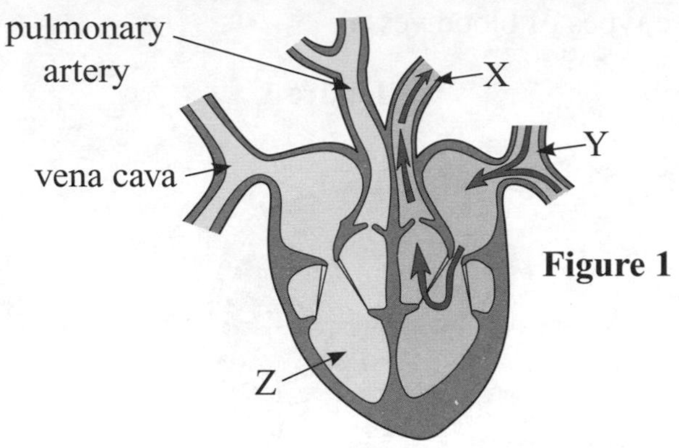

Figure 1

1.1 Name the parts of the heart labelled X, Y and Z in **Figure 1**.

X .... Y .... Z ....

[3]

1.2 Draw arrows on **Figure 1** to show the direction of blood flow through the right side of the heart.

[1]

1.3 Explain why the human circulatory system is described as a 'double circulatory system'.

....

....

....

....

....

[3]

[Total 7 marks]

2 The heart beats to circulate blood around the body.

2.1 Describe how the heartbeat is controlled.

....

....

[2]

2.2 Atrial fibrillation is a condition where the heartbeat is irregular. It is caused by problems with the heart's ability to control its own beat. Suggest how atrial fibrillation could be treated.

....

....

....

[2]

[Total 4 marks]

Circulatory System — Blood Vessels

1 Blood is carried around the body in blood vessels.
Different types of blood vessel perform different functions.

Figure 1 shows the three types of blood vessel.

Figure 1

A B C

1.1 Which of these blood vessels, A, B or C is an artery? Tick **one** box.

☐ A ☐ B ☐ C

[1]

1.2 The blood in arteries flows under high pressure.
Explain how arteries are adapted to perform their function.

....

....

....

[2]

1.3 Name the type of blood vessel that has valves.

....

[1]

1.4 Why does the blood vessel named in 1.3 have valves?

....

....

[1]

1.5 Explain why the walls of capillaries are only one cell thick.
Refer to their function in your answer.

....

....

....

....

....

[2]

[Total 7 marks]

Circulatory System — Blood

1 Blood is made up of several different components, including white blood cells, red blood cells and platelets. Grade 7-9

1.1 Some diseases affect the body's ability to produce enough white blood cells.
Suggest why people with these diseases are more likely to experience frequent infections.

....
[1]

1.2 Explain how white blood cells are adapted to perform their function.

....

....

....

....
[3]

1.3 Red blood cells carry oxygen from the lungs to other tissues in the body.
Explain how red blood cells are adapted for their function.

....

....

....

....

....
[3]

The components of blood can be separated by spinning them at high speed.
Figure 1 shows a tube of blood that has been separated in this way.

Figure 1

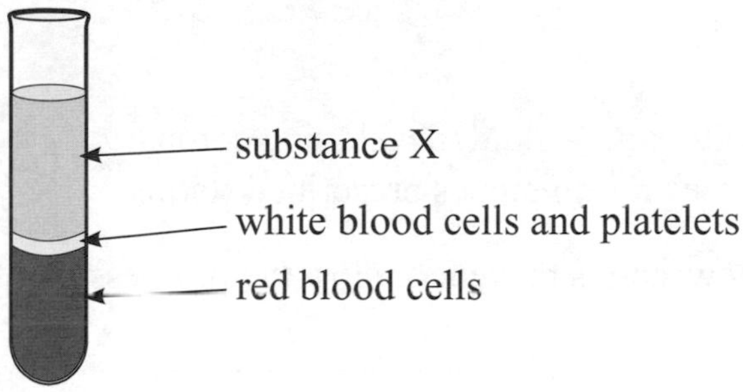

substance X
white blood cells and platelets
red blood cells

1.4 Identify the substance labelled X.

....
[1]

1.5 A scientist analysing the blood sample found that it had a lower than normal concentration of platelets. Describe the structure and function of platelets.

....

....
[2]

[Total 10 marks]

Cardiovascular Disease

Warm-Up

Use the correct words to fill in the gaps in the passage. Not all of them will be used.

pulmonary vein   blood vessels   asthma   aorta   coronary heart disease
coronary arteries   fatty acids   toxins   fatty material   cystic fibrosis

Cardiovascular disease is a term used to describe diseases of the ....

and heart. .... is an example of a cardiovascular disease.

It is caused by narrowing of the .... due to the build-up of

.... on the inside wall.

1 The coronary arteries surround the heart.
A patient has a blockage of fatty material in a coronary artery. (Grade 6-7)

1.1 Explain why a blockage in the coronary arteries could cause damage to the patient's heart muscle.

....

....

[2]

1.2 Suggest and describe a method of treatment that a doctor might recommend to the patient.

....

....

....

[2]

[Total 4 marks]

2 Patients with, or at risk of, developing coronary heart disease are sometimes prescribed statins. (Grade 6-7)

2.1 Explain how statins prevent or slow the progression of coronary heart disease.

....

....

....

[2]

2.2 A patient is offered statins. Suggest **one** reason why he may not wish to take them.

....

....

[1]

[Total 3 marks]

3 Doctors were assessing the heart of a patient recovering from a serious heart infection. 

3.1 They found that one of the valves in the heart had become leaky.
Suggest the effects this might have on blood flow through the heart and around the body.

....

....

....
[2]

3.2 Describe **one** other way that a valve might be faulty.

....
[1]

Surgeons decided to replace the faulty valve with a replacement biological valve.

3.3 What is a biological valve?

....
[1]

3.4 A mechanical valve is sometimes used in transplants instead of a biological valve.
What is a mechanical valve?

....
[1]

A second patient at the same hospital needed a heart transplant.
Heart transplants can use donor hearts or artificial hearts.

3.5 Artificial hearts are rarely used as a permanent fix.
Suggest when they are most likely to be used instead of a natural donor heart.

....

....

....
[2]

3.6 Suggest **one** advantage and **one** disadvantage of using a natural donor heart rather than an artificial heart in heart transplant operations.

Advantage ....

....

Disadvantage ....

....
[2]

[Total 9 marks]

Health and Disease

1 Ill health is often caused by communicable and non-communicable diseases. Grade 4-6

1.1 What is meant by the term 'communicable disease'?

....

....

[1]

1.2 List **two** factors other than disease that can cause ill health.

....

....

[2]

[Total 3 marks]

2 **Figure 1** and **Table 1** show the number of employees in five different rooms in a large office building who have had at least one common cold in the last 12 months. Grade 6-7

Figure 1

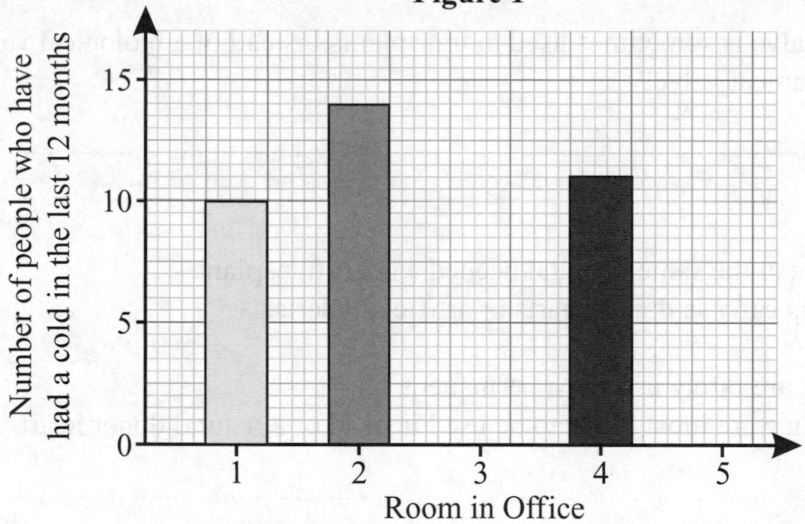

Table 1

	Room 1	Room 2	Room 3	Room 4	Room 5	**Total**
Number of people who have had a cold in the last 12 months	10		12	11		**60**

2.1 Complete **Figure 1** and **Table 1**.

[4]

2.2 Some people have a defective immune system. Explain what effect this could have on the likelihood of a person contracting a communicable disease like the common cold.

....

....

....

[2]

[Total 6 marks]

Risk Factors for Non-Communicable Diseases

1 Substances in a person's environment can be risk factors for certain diseases. Grade 4-6

1.1 What is meant by a risk factor for a disease?

....

....

....

[1]

1.2 Other than substances in the environment, state **two** other types of risk factor.

....

....

[2]

1.3 Obesity is a risk factor for many different diseases.
Name **one** disease that obesity is a risk factor for.

....

[1]

[Total 4 marks]

2 A patient has been diagnosed with cardiovascular disease. Grade 4-6

2.1 Give **two** risk factors that might have contributed to her developing cardiovascular disease.

....

....

[2]

2.2 Suggest **two** reasons why non-communicable diseases can be financially costly.

....

....

....

....

[2]

[Total 4 marks]

Exam Practice Tip

Remember that risk factors are identified by scientists looking for correlations in data, and they don't always directly cause a disease. Sometimes, they are just related to another factor that does. It's rarely a single risk factor that leads to someone developing a disease — diseases are often caused when multiple risk factors interact with each other.

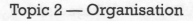

Cancer

Warm-Up

Tumours can be benign or malignant. Draw lines to match the types of tumour on the left with each characteristic on the right that applies to them.

Malignant Tumours

Benign Tumours

Are cancerous

Are not cancerous

Can invade neighbouring tissues

1 Some types of tumour are cancerous. Grade 4-6

1.1 What do tumours result from?
Tick **one** box.

- [ ] Rapid cell death
- [ ] Slow cell division
- [ ] Lack of cell division
- [ ] Uncontrolled cell division

[1]

1.2 There are many known lifestyle-related risk factors for cancer. However, not all risk factors for cancer are related to lifestyle. Give **one** other type of risk factor for cancer.

....

[1]

[Total 2 marks]

2 Doctors found a tumour in the left lung of a patient.
They were concerned that the patient was at risk of developing secondary tumours. Grade 6-7

2.1 Was the tumour in the patient's lung malignant or benign?

....

[1]

2.2 Explain how a secondary tumour forms.

....

....

....

[2]

[Total 3 marks]

Plant Cell Organisation

1 The roots, stem and leaves are involved in the transport of substances around a plant. Grade 4-6

1.1 What do the roots, stem and leaves of a plant form? Tick **one** box.

☐ A tissue ☐ An organ system ☐ An organ ☐ A tissue system

[1]

1.2 Name **two** substances that are transported around a plant in the xylem.

....

[2]

[Total 3 marks]

2 Plants have many types of tissue, including meristem tissue. Grade 4-6

2.1 Name **two** sites in a plant where you would find meristem tissue.

....

....

[2]

2.2 Give **one** reason why meristem tissue is important throughout the life of the plant.

....

[1]

[Total 3 marks]

3 **Figure 1** shows a transverse section of a leaf. Grade 6-7

Figure 1

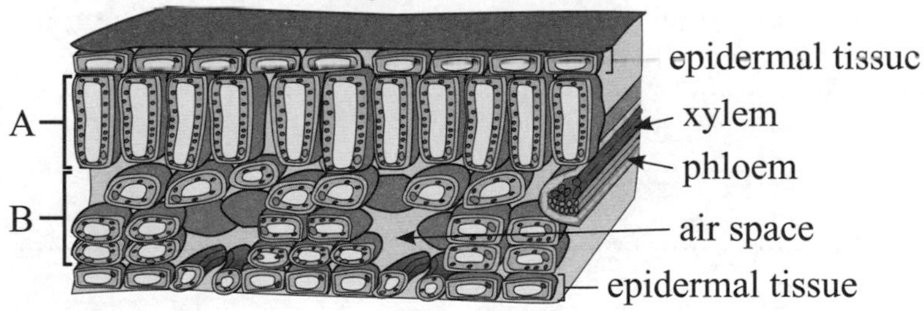

3.1 Name the tissues labelled A and B.

A .... B ....

[2]

3.2 Explain how the tissue labelled A is adapted for the function of photosynthesis.

....

....

....

[2]

3.3 What is the function of the air spaces?

....

[1]

[Total 5 marks]

☐ ☐ ☐

Transpiration and Translocation

Warm-Up

Use the words below to correctly fill in the gaps in the passage.
You don't have to use every word, but each word can only be used once.

leaves phloem translocation mineral ions condensation evaporation
roots perspiration xylem transpiration sugars guard cells stem

The process by which water is lost from a plant is called ....

It is caused by the .... and diffusion of water from a plant's surface,

most often from the .... Another process, called ....,

is the transport of .... from where they're made in the leaves to the rest

of the plant via the .... vessels.

1* Describe how xylem tissue and phloem tissue work to:

- supply water and mineral ions to all parts of a plant,
- transport dissolved sugars around a plant.

Grade 7-9

Include details of the **structure** and **function** of the tissues involved.

....

....

....

....

....

....

....

....

....

....

....

....

....

[Total 6 marks]

Exam Practice Tip

It can be tricky to remember which is which when thinking of xylem and phloem. They're both pretty similar and they've both got weird names. If you're struggling to remember, keep practicing this question. It will soon stick in your head.

Transpiration and Stomata

1 **Figure 1** is a drawing of a magnified image of part of the surface of a leaf

Figure 1

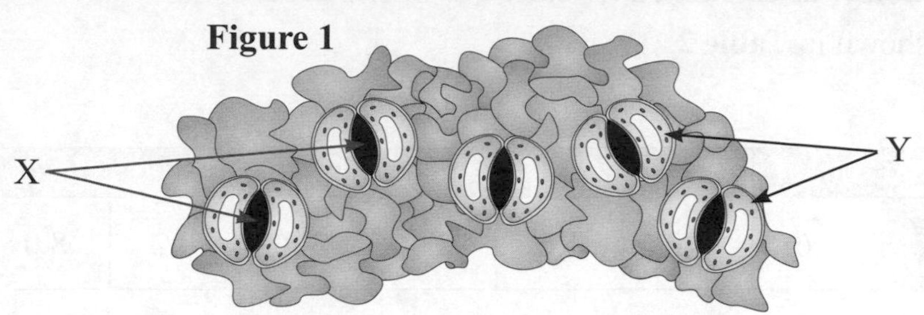

1.1 Name the structures labelled X and the cells labelled Y in **Figure 1**.

X .... Y ....

[2]

1.2 What is the function of the cells labelled Y?

....

....

....

[2]

[Total 4 marks]

2 **Table 1** shows the diameter of ten stomata from each of two leaves, A and B.

Grade 6-7

Table 1

Diameter of stomata (μm)	
Leaf A	Leaf B
25.2, 20.1, 18.7, 17.9, 19.1, 19.3, 22.0, 23.1, 21.8, 20.3	14.7, 12.8, 14.1, 13.2, 12.9, 11.9, 12.1, 13.4, 10.9, 11.7

2.1 Calculate the mean width of the stomata for each leaf.

Leaf A = ... μm Leaf B = ... μm

[2]

2.2 Leaves A and B are from the same species. Which leaf do you think had its stomatal measurements taken in lower light intensity?

....

[1]

2.3 Explain your answer to 2.2.

....

....

....

[2]

[Total 5 marks]

3 An investigation was carried out to assess the rate of water uptake by a plant over a 16-hour period.

A potometer was set up and readings were taken every two hours between 00:00 and 16:00.
The rate in cm^3/hour was calculated for each two-hour period.
The results are shown in **Table 2**.

Table 2

Time of day	00:00	02:00	04:00	06:00	08:00	10:00	12:00	14:00	16:00
Rate of water uptake (cm^3/hour)	2.6	1.0	1.6	2.0	3.8	6.2	8.0	10.2	7.6

3.1 Complete **Figure 2** using the data displayed in **Table 2**.
- Select a suitable scale and label for the y-axis
- Plot the rate of water uptake in cm^3/hour for all the times given in **Table 2**
- Join the points with straight lines

Figure 2

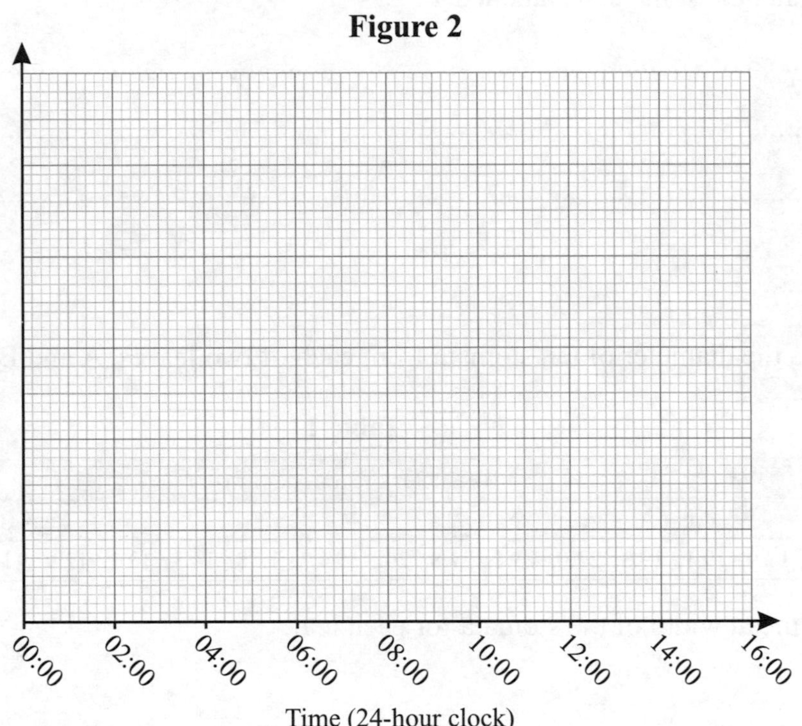

[4]

Use your graph to answer 3.2 and 3.3 below.

3.2 Use the graph to estimate the rate of water uptake at 09:00.

....

[1]

3.3 How much did the rate of water uptake increase between 06:00 and 11:00?

....

[1]

3.4 Suggest **two** environmental changes that could have caused the change in water uptake between 06:00 and 14:00.

....

....

[2]

[Total 8 marks]

Communicable Disease

1 Viruses and bacteria can both reproduce inside the human body. Grade 4-6

1.1 Which of the following sentences is correct? Tick **one** box.

☐ Both bacteria and viruses can reproduce quickly in the body.

☐ Bacteria reproduce quickly in the body, but viruses reproduce slowly.

☐ Viruses reproduce quickly in the body, but bacteria reproduce slowly.

☐ Both bacteria and viruses reproduce slowly in the body.

[1]

1.2 Viruses reproduce inside cells. Describe what problem this can cause for the cells.

....

[1]

[Total 2 marks]

2* **Figure 1** shows a housefly. Houseflies are vectors because they can transmit disease to humans. Grade 7-9

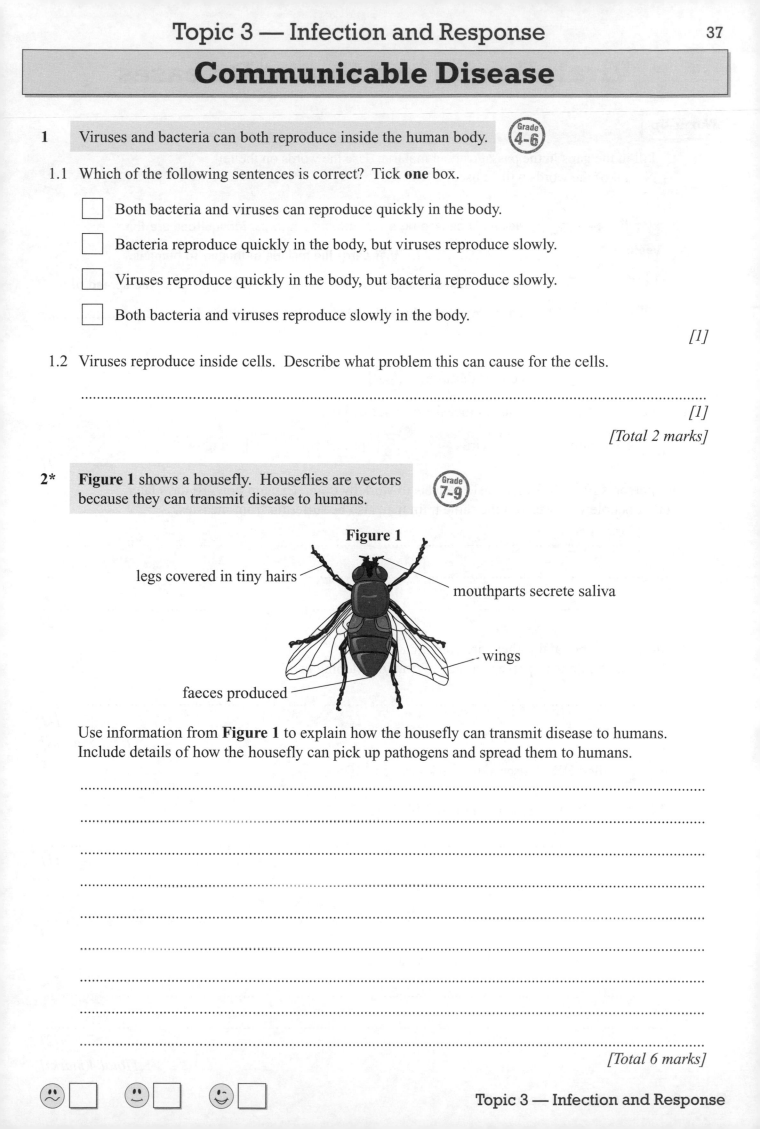

Figure 1

Use information from **Figure 1** to explain how the housefly can transmit disease to humans. Include details of how the housefly can pick up pathogens and spread them to humans.

....

....

....

....

....

....

....

....

....

[Total 6 marks]

Viral, Fungal and Protist Diseases

Warm-Up

Fill in the gaps in the passage about malaria. Use the words on the left.
Not all of the words will be used.

protist breeding vectors fungi fever bacterium virus

Malaria is caused by a .... Mosquitoes are the .... that carry the malaria pathogen to humans.

Malaria causes repeating episodes of .... The spread of malaria can be reduced by stopping the mosquitoes from ....

1 Measles is a highly infectious disease. Grade 4-6

1.1 What type of pathogen causes measles? Tick **one** box.

☐ bacterium ☐ virus ☐ protist ☐ fungus

[1]

1.2 A person suffering from measles travels to work by train. Explain how, 10 days later, other people who were on the same train may also be suffering from measles.

....

....

....

[3]

1.3 Measles can be fatal if there are complications.
What can be done to prevent someone from developing measles?

....

[1]

[Total 5 marks]

2 A virus called HIV causes a disease known as AIDS. Grade 4-6

2.1 What type of drug can be used to control HIV?

....

[1]

2.2 What system in the body does HIV attack?

....

[1]

2.3 State **two** ways in which HIV can be spread.

....

....

[2]

[Total 4 marks]

3 The tobacco mosaic virus (TMV) is a widespread plant pathogen affecting many species of plants. Grade 6-7

3.1 Name **one** species of plant that is often attacked by the tobacco mosaic virus.

....
[1]

3.2 Describe the appearance of a plant with TMV.

....
[1]

3.3 Outline why a plant affected by TMV cannot grow properly.

....

....
[1]

3.4 **Table 1** shows the mean diameter and mass of fruits from 100 healthy plants and 100 plants infected with TMV.

Table 1

	Healthy plants	Plants with TMV
Mean diameter of fruit (mm)	50	35
Mean mass of fruit (g)	95	65

Describe the effect of TMV on the diameter and mass of fruit produced in the infected plants compared to the healthy plants.

....

....

....
[2]

[Total 5 marks]

4 Rose black spot is a disease that can affect rose plants. Grade 6-7

4.1 Describe the appearance of leaves that are infected with rose black spot and state what happens to these leaves.

....

....
[3]

4.2 A gardener notices that one of her rose plants is infected with rose black spot. She is worried about the rest of her rose plants becoming infected with the fungus. Why are the other rose plants in her garden at risk from being infected with the disease?

....
[1]

4.3 Rose black spot can be treated by removing and destroying the infected leaves and treating the rest of the plant with a fungicide. Suggest why it is important to destroy the removed leaves.

....

....
[1]

[Total 5 marks]

Bacterial Diseases and Preventing Disease

1 Salmonella food poisoning in humans is caused by a bacterium. (Grade 4-6)

1.1 List **two** symptoms of *Salmonella* food poisoning.

....

....

[2]

1.2 What does the *Salmonella* bacterium produce that causes these symptoms?

....

[1]

1.3 In the UK, poultry are vaccinated against the bacterium that causes food poisoning.
Explain why it is necessary to vaccinate poultry.

....

....

....

[2]

1.4 Suggest **one** way that a person suffering from *Salmonella* food poisoning can prevent passing the disease on to someone else.

....

[1]

[Total 6 marks]

2 Gonorrhoea is a disease that can affect both men and women. (Grade 4-6)

2.1 How is gonorrhoea spread from person to person?

....

[1]

2.2 State **two** symptoms of the disease in women.

....

....

[2]

2.3 Name the antibiotic that was originally used to treat people infected with gonorrhoea.

....

[1]

2.4 Why is the antibiotic in part 2.3 no longer effective against the bacterium that causes gonorrhoea?

....

[1]

2.5 Name **one** barrier method of contraception that prevents the spread of gonorrhoea.

....

[1]

[Total 6 marks]

Fighting Disease

1 The body has many features that it can use to protect itself against pathogens. Grade 4-6

1.1 Describe how the skin helps to defend the body against pathogens.

....

....

[2]

1.2 How do structures in the nose help to defend the body against the entry of pathogens?

....

....

[1]

[Total 3 marks]

2* Describe how the human body works to defend itself against pathogens that have entered the body. Include details of the body's defences and the role of the immune system. Grade 6-7

....

....

....

....

....

....

....

....

....

....

....

....

[Total 6 marks]

Exam Practice Tip

Think carefully about 6 mark questions like the one on this page. Don't just start scribbling everything you know about the topic. Stop and think first — work out what the question is wanting you to write about, and then make sure you write enough points to bag yourself as many marks as possible. Good job you've got some practice on this page.

Fighting Disease — Vaccination

1 Children are often vaccinated against measles. Grade 4-6

1.1 What is injected into the body during a vaccination?

[1]

1.2 Describe what happens when a vaccine is injected into the body. Tick **one** box.

☐ Red blood cells are stimulated to produce antibodies.

☐ White blood cells are stimulated to produce antibiotics.

☐ Red blood cells are stimulated to produce antibiotics.

☐ White blood cells are stimulated to produce antibodies.

[1]

[Total 2 marks]

2 People can be vaccinated against a large number of diseases. Grade 6-7

2.1 If the mumps pathogen enters the body of someone who has had the mumps vaccination, why would they be unlikely to become ill with mumps?

[1]

2.2 A large proportion of a population is vaccinated against a particular pathogen. Suggest why the spread of the pathogen will be very much reduced.

[2]

[Total 3 marks]

3 When visiting some other countries, it is recommended that travellers are vaccinated against some of the serious diseases found in those countries. Grade 7-9

3.1 If a traveller planned to visit a country where there had been a recent outbreak of the communicable disease cholera, they might get vaccinated against cholera before they travelled. Give **two** reasons why this would be beneficial.

[2]

3.2 Some countries insist that travellers have had particular vaccinations before they are allowed to enter the country. Suggest why.

[1]

[Total 3 marks]

Fighting Disease — Drugs

1 There are many different types of drugs with different functions. (Grade 6-7)

1.1 Explain why it is difficult to develop drugs to kill viruses.

....

....
[2]

1.2 Many people suffer from sore throats caused by bacteria. Other than an antibiotic, name a type of drug that could be used to ease the symptoms of the infection.

....
[1]

1.3 Explain why the type of drug named in 1.2 would not be able to cure the bacterial infection.

....
[1]

[Total 4 marks]

2 A hospital records the number of cases of infections that are caused by antibiotic-resistant bacteria each year. The figures for three years are shown in **Table 1**. (Grade 6-7)

Table 1

Year	2013	2014	2015
No. of infections	84	102	153

2.1 What is meant by antibiotic-resistant bacteria?

....

....
[1]

2.2 Describe the trend shown in **Table 1**.

....

....
[1]

2.3 Use **Table 1** to calculate the percentage change in antibiotic resistant infections between 2013 and 2015.

.... %

[2]

[Total 4 marks]

Exam Practice Tip

When you're answering an exam question about drugs and disease, think very carefully about whether the drug kills the pathogens causing the disease (and so cures it), or whether it just helps to make the symptoms of the disease better.

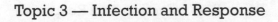

Developing Drugs

1 New drugs have to be tested and trialled before they can be used. Grade 4-6

1.1 List **three** things drugs must be tested for, to ensure they are safe and effective.

....

....

....
[3]

1.2 Which of the following is preclinical testing carried out on? Tick **one** box.

☐ healthy human volunteers ☐ patients in a hospital

☐ cells, tissues and dead animals ☐ cells, tissues and live animals

[1]

[Total 4 marks]

2 Clinical trials are always carried out on healthy volunteers before patients. Grade 7-9

2.1 Suggest why very low doses of the drug are given at the start of clinical trials.

....
[1]

2.2 Clinical trials are often double-blind. Explain what would happen in a double-blind clinical trial.

....

....

....
[3]

2.3 Suggest why clinical trials are carried out using a double-blind method.

....

....
[1]

The final results of clinical trials cannot be published until they have been checked by other scientists. Such checking is often referred to as peer review.

2.4 Suggest why peer review is important in these trials.

....
[1]

2.5 Suggest why it is important that the scientists who carry out the peer review have no links to the people who carried out the original trials.

....

....
[2]

[Total 8 marks]

Monoclonal Antibodies

Warm-Up

Circle the correct underlined words below, so that each sentence is correct.

Monoclonal antibodies are made by lymphocytes/pathogens. They can be used to locate particular protein antigens in a blood sample by being bound to a fluorescent dye/drug. The monoclonal antibodies will attach to/dye the protein antigens and can be detected.

1 **Figure 1** shows a monoclonal antibody. It will attach to a particular antigen. 

Figure 1

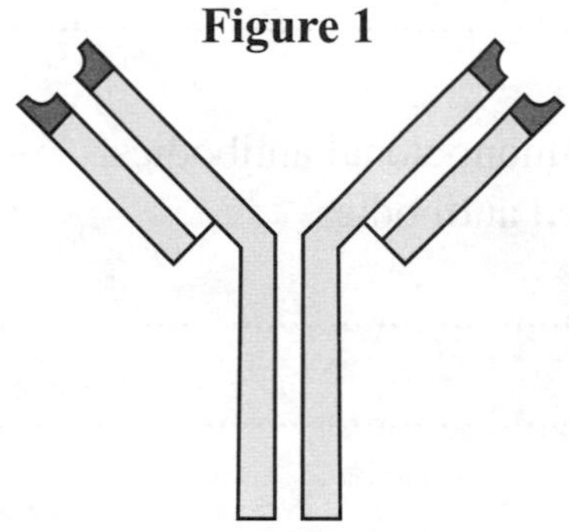

1.1 On **Figure 1**, circle where the antibody will bind to the antigen.

[1]

Figure 2 shows some antigens.

Figure 2

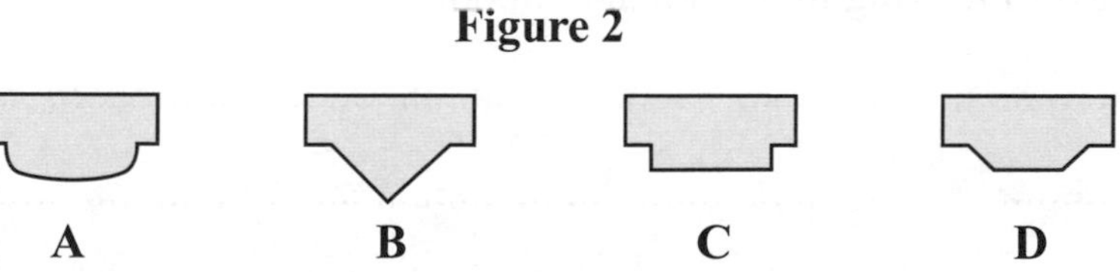

1.2 Which antigen will the antibody in **Figure 1** attach to? Tick **one** box.

☐ Antigen **A** ☐ Antigen **B** ☐ Antigen **C** ☐ Antigen **D**

[1]

[Total 2 marks]

2 Monoclonal antibodies have lots of uses. 

What are monoclonal antibodies?

....

....

....

[Total 2 marks]

3 Monoclonal antibodies can be used to treat cancer. Grade 6-7

3.1 Explain how monoclonal antibodies can be used to treat cancer.

.....

[4]

3.2 When monoclonal antibodies were first discovered, it was thought that they would be used more widely in medical treatments than they actually are. Explain why this has not been the case.

.....

[1]

3.3 Treating cancer is just one use of monoclonal antibodies.
Give **two** more uses of monoclonal antibodies.

.....

[2]

[Total 7 marks]

4* Monoclonal antibodies are engineered by scientists. Grade 7-9

Describe the process of making monoclonal antibodies.

.....

[Total 6 marks]

Plant Diseases and Defences

1 Plants can be infected by a range of different types of pathogens. (Grade 4-6)

1.1 Name a virus that plants can be infected with.

....

[1]

1.2 Name a fungus that can cause disease in plants.

....

[1]

1.3 Insects can also affect plants. Name **one** insect that can cause damage to plants.

....

[1]

[Total 3 marks]

2 A gardener notices that one of his plants has spots on its leaves. He thinks that this might be a sign of a disease. (Grade 6-7)

2.1 List **four** other signs that a plant has a disease.

....

....

....

....

[4]

2.2 The gardener tries using a gardening manual to identify the disease.
Suggest **two** other ways that the gardener could identify the disease that has infected his plant.

....

....

[2]

2.3 A gardener counts that he has 42 plants in his garden.
Table 1 shows the number of those plants infected by a pathogen.

Table 1

	Infected by fungus	Infected by virus
No. of plants	12	6

Calculate the percentage of plants that have a viral infection.

.... %

[2]

[Total 8 marks]

3 Plants have many ways of defending themselves. (Grade 6-7)

3.1 Suggest how having cellulose cell walls protects a plant from microorganisms.

....

[1]

3.2 Give **two** types of chemical that plants can produce for defence.
For each type of chemical, explain how it helps the plant to survive.

....

....

....

....

[4]

3.3 State **two** types of mechanical adaptation that help plants to protect themselves.
For each one, explain how it works.

....

....

....

....

[4]

[Total 9 marks]

4 Plants need a range of minerals ions in order to remain healthy. (Grade 6-7)

4.1 A plant has stunted growth. Suggest what mineral it is deficient in.

....

[1]

4.2 Explain why a deficiency in the mineral you named in 4.1 causes stunted growth in plants.

....

....

[2]

4.3 Other than the mineral you named in 4.1, name another mineral ion that plants require for healthy growth.

....

[1]

4.4 Explain how a deficiency in the ion you named in 4.3 affects a plant.

....

....

....

[2]

[Total 6 marks]

Photosynthesis and Limiting Factors

1 Photosynthesis is where energy is transferred to plants and used to make glucose. Grade 4-6

1.1 What is the source of energy for photosynthesis?

....

[1]

1.2 Complete the following word equation for photosynthesis.

.... + water → glucose + ...

[2]

Plants use the glucose they produce in lots of different ways, including to make a substance to strengthen cell walls.

1.3 Name the substance that plants use to strengthen cell walls.

....

[1]

1.4 Give **two** other ways that plants use the glucose produced during photosynthesis.

....

....

....

....

[2]

[Total 6 marks]

2 Photosynthesis is an endothermic reaction. Various factors affect its rate. Grade 6-7

2.1 Explain what is meant by an endothermic reaction.

....

[1]

2.2 Which of the following factors does not affect the rate of photosynthesis? Tick **one** box.

☐ carbon dioxide concentration ☐ nitrate concentration ☐ light intensity ☐ temperature

[1]

2.3 A lack of magnesium can cause chloroplasts not to make enough chlorophyll. Explain what effect this would have on the rate of photosynthesis of a plant.

....

....

[2]

[Total 4 marks]

The Rate of Photosynthesis

Warm-Up

Choose from the words below to complete the sentences explaining how temperature affects the rate of photosynthesis, as shown in the graph. Some words may not be used at all.

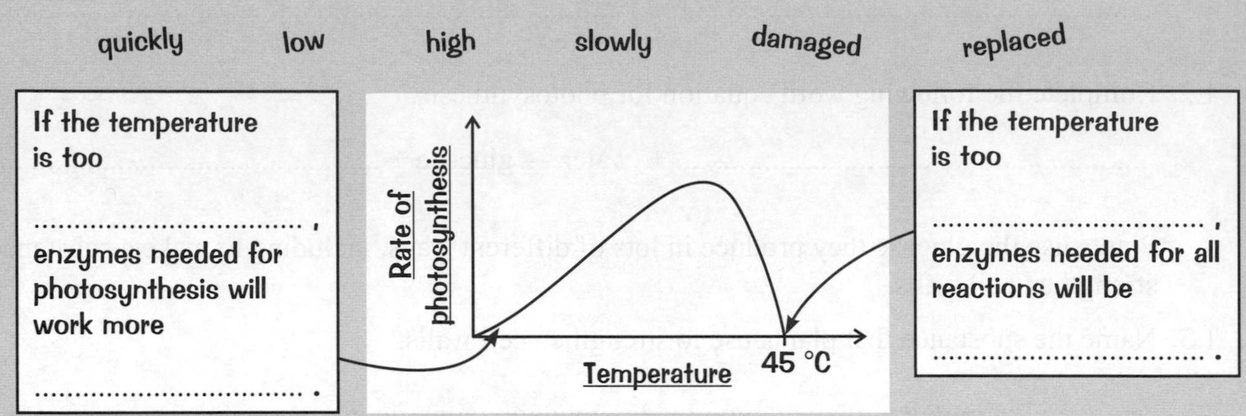

quickly
low
high
slowly
damaged
replaced
If the temperature is too .... , enzymes needed for photosynthesis will work more ....
Rate of photosynthesis
Temperature
45 °C
If the temperature is too .... , enzymes needed for all reactions will be ....

1 **Figure 1** shows a greenhouse. Greenhouses are used to create the ideal conditions for photosynthesis.

Figure 1

1.1 Suggest **two** ways that a farmer could improve the conditions for photosynthesis in a greenhouse. For each of the ways, explain how it affects the rate of photosynthesis.

Improvement ....

Explanation ....

....

Improvement ....

Explanation ....

....

[4]

1.2 Creating the ideal conditions in a greenhouse costs money. Explain why it may still be beneficial for the farmer to do this.

....

....

[2]

[Total 6 marks]

2 A student carried out an experiment to investigate the effect of changing the concentration of carbon dioxide on the rate of photosynthesis in a green plant. The results were plotted on the graph shown in **Figure 2**.

Figure 2

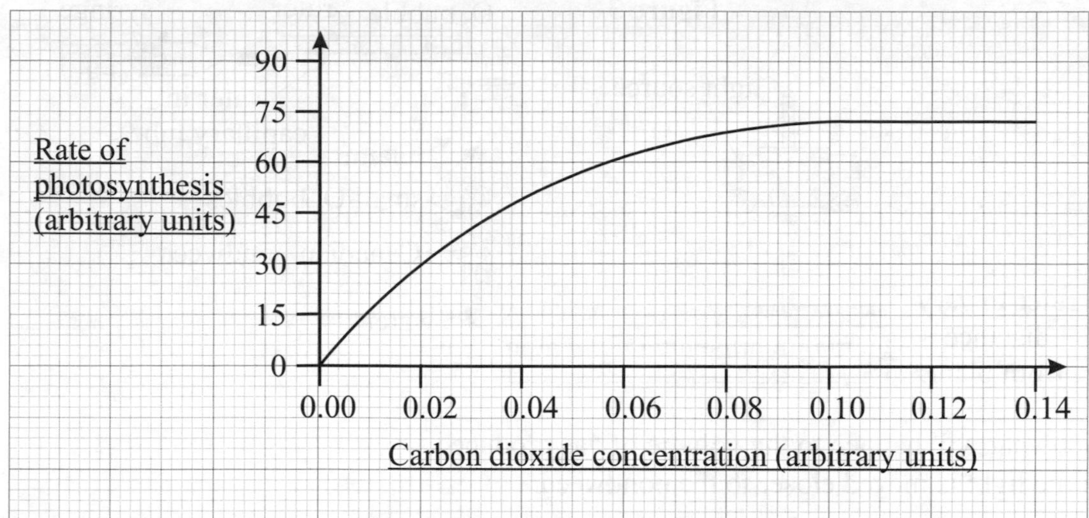

2.1 Describe the trend shown in the graph.

[2]

2.2 At a certain point, the CO_2 concentration is no longer limiting the rate of photosynthesis. Suggest **two** factors that could be limiting the rate at this point.

[2]

2.3 In the space below, sketch a graph to show how light intensity affects the rate of photosynthesis.

[2]

[Total 6 marks]

PRACTICAL

3 A student was investigating the effect of light intensity on the rate of photosynthesis in a water plant. She set up the experiment as shown in **Figure 3**.

Figure 3

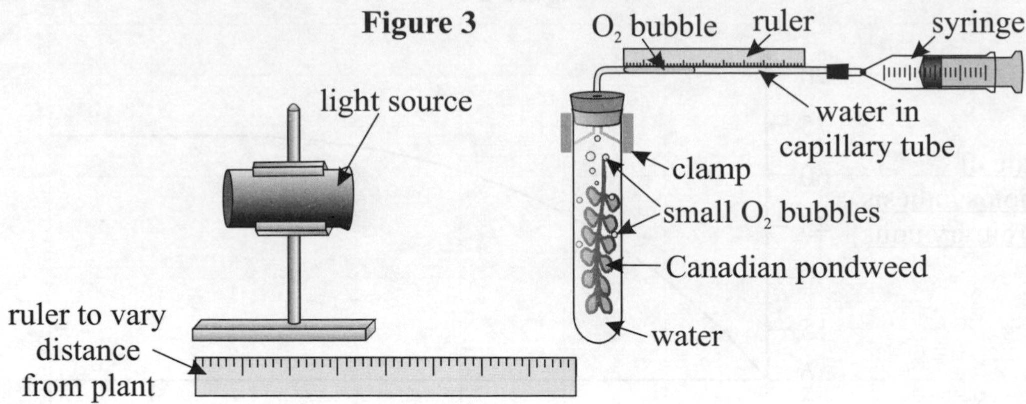

3.1 Predict what will happen to the volume of gas produced when the light is moved closer to the pondweed.

....

[1]

3.2 The formula 1/distance^2 can be used as a measure of light intensity. It's called the inverse square law. Use the inverse square law to calculate the light intensity when the lamp is 20 cm from the pondweed.

light intensity = ... arbitrary units

[2]

3.3* Suggest how the student could adapt the experiment shown in **Figure 3** to investigate the effect of temperature on the rate of photosynthesis. Include details of the variables that should be controlled.

....

....

....

....

....

....

....

....

[6]

[Total 9 marks]

Exam Practice Tip

There's a lot to learn about limiting factors and the rate of photosynthesis. It's a good idea to practise drawing the graphs to show the effect of light intensity, carbon dioxide concentration and temperature on the rate of photosynthesis. You should also make sure you're able to interpret the graphs too, including those with more than one factor involved.

Respiration and Metabolism

1 Respiration is a reaction carried out by all living organisms. It transfers energy from an organism's food to their cells. Grade 6-7

1.1 Name the type of reaction where energy is transferred to the environment.

[1]

Figure 1 shows a gull.

Figure 1

1.2 Give **three** examples of how a gull uses the energy transferred by respiration.

[3]

[Total 4 marks]

2 Metabolism is the sum of all of the reactions that happen in a cell or the body. Metabolism includes reactions that make molecules. Grade 6-7

2.1 Some metabolic reactions involve using glucose molecules to make other molecules.
Name a molecule made from glucose in plants, and a molecule made from glucose in animals.

Plants

Animals

[2]

2.2 Describe the components of a lipid molecule.

[2]

2.3 Briefly describe how protein molecules are formed.

[2]

2.4 Metabolism also involves breaking down molecules.
What is excess protein broken down to produce?

[1]

[Total 7 marks]

Aerobic and Anaerobic Respiration

Warm-Up

Draw lines to match up each process on the left with its correct description on the right.

Aerobic respiration

Anaerobic respiration

Fermentation

Respiration without oxygen.

Respiration using oxygen.

1 An experiment was set up using two sealed beakers, each with a carbon dioxide monitor attached. The set up is shown in **Figure 1**.

Figure 1

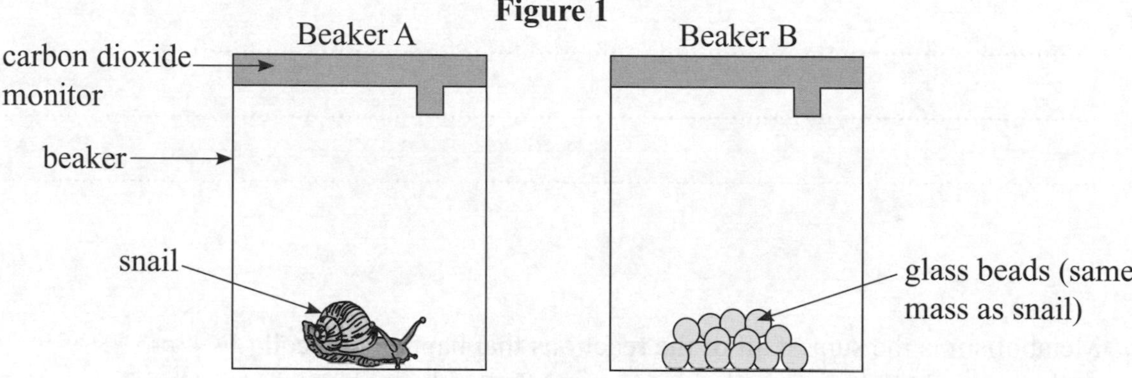

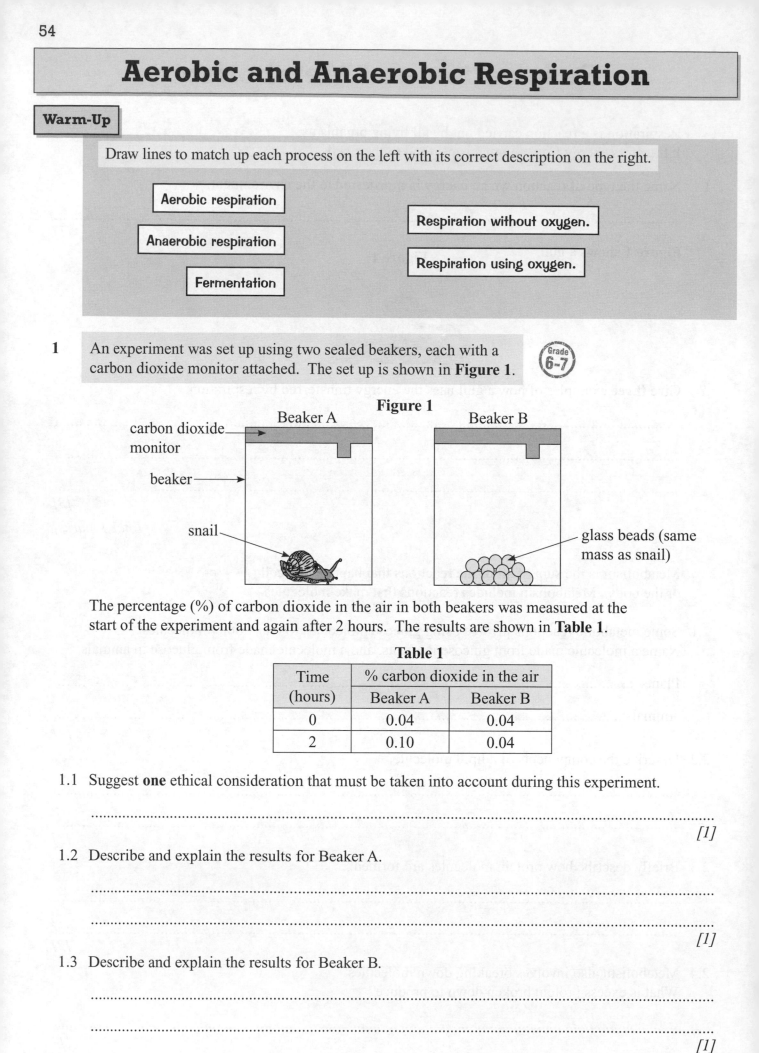

The percentage (%) of carbon dioxide in the air in both beakers was measured at the start of the experiment and again after 2 hours. The results are shown in **Table 1**.

Table 1

Time (hours)	% carbon dioxide in the air	
	Beaker A	Beaker B
0	0.04	0.04
2	0.10	0.04

1.1 Suggest **one** ethical consideration that must be taken into account during this experiment.

....
[1]

1.2 Describe and explain the results for Beaker A.

....

....
[1]

1.3 Describe and explain the results for Beaker B.

....

....
[1]

1.4 What would have happened to the level of oxygen in Beaker A after two hours?
Explain your answer.

....

....
[2]

1.5 In the experiment, Beaker B was set up in the same way as Beaker A but with glass beads instead of a snail. Suggest why Beaker B is used in the experiment.

....

....
[1]

[Total 6 marks]

2 *S. cerevisiae* is a type of yeast. It carries out fermentation. (Grade 6-7)

2.1 Complete the following word equation for fermentation.

.... → ethanol + carbon dioxide
[1]

2.2 For each of the products of fermentation in yeast, outline **one** industrial use.

Product ....

Industrial use ....

Product ....

Industrial use ....
[2]

[Total 3 marks]

3 **Figure 2** shows a muscle cell.
Compare aerobic respiration and anaerobic respiration in muscle cells. (Grade 6-7)

Figure 2

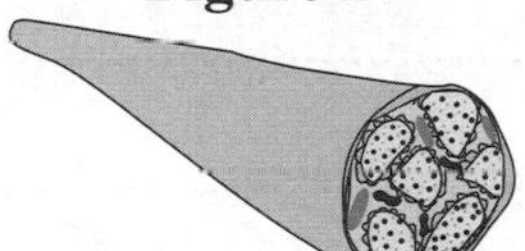

....

....

....

....

....

....

[Total 3 marks]

Exercise

Warm-Up

Choose from the words on the right to complete the sentences about exercise. Some words may not be used at all.

carbon dioxide oxygen aerobically lactic acid urea oxygen debt muscles

During exercise your .... may respire aerobically resulting in an .... This is the amount of extra .... your body needs to react with the build up of .... and remove it from cells.

1 A student was investigating the effect of exercise on his own breathing rate. He counted his number of breaths per minute before, during and after a period of exercise. He repeated his experiment another two times. The results are shown in **Table 1** below.

Table 1

	Breathing rate (number of breaths per minute)			
	Before exercise	During exercise	One minute after exercise	Five minutes after exercise
	11	16	15	12
	12	15	14	11
	11	15	14	12
Mean	11	15	14	

1.1 Calculate the mean breathing rate five minutes after exercise.

Mean = ... breaths per minute

[1]

1.2 Describe and explain how the student's breathing rate changed during exercise compared to before exercise, as shown in **Table 1**.

....

....

....

[3]

1.3 Describe and explain how the student's breathing rate changed after exercise compared to during exercise, as shown in **Table 1**.

....

....

....

....

[5]

1.4 Give **two** other variables that the student could have measured which would have shown the same trend as breathing rate during exercise.

....

....

[2]

[Total 11 marks]

2 **Figure 1** below shows the amount of lactic acid in an athlete's body before, during and after some vigorous exercise.

Grade 7-9

Figure 1

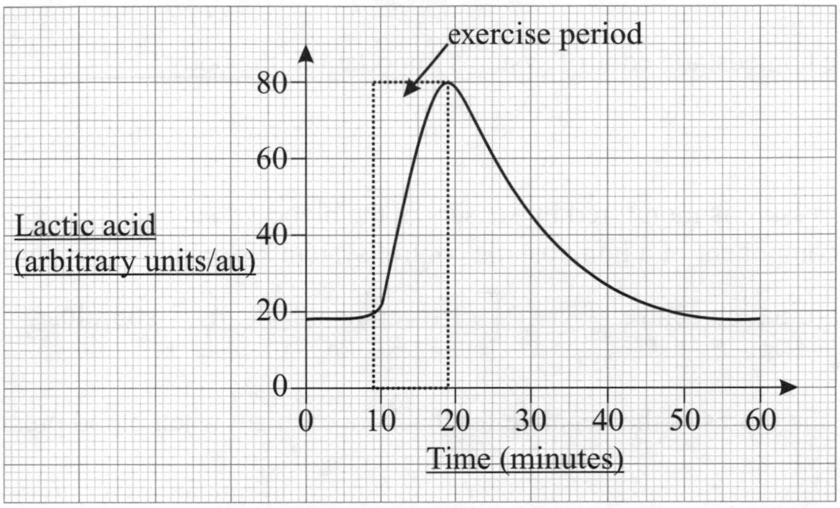

exercise period

Lactic acid (arbitrary units/au)

80
60
40
20
0

0 10 20 30 40 50 60

Time (minutes)

2.1 The amount of lactic acid increases from 20 au to 80 au across the duration of the exercise period. Work out the percentage change in lactic acid during this time.

Percentage change = ...%

[2]

2.2 Explain why there was an increase in lactic acid during the exercise period.

....

....

....

[3]

2.3 What physical effects does a long period of vigorous exercise have on muscles?

....

....

[2]

2.4 The amount of lactic acid in the athlete's body decreases between 20 and 60 minutes. Explain how this lactic acid is cleared from the body.

....

....

[2]

[Total 9 marks]

Homeostasis

1 Human blood pressure is maintained by a homeostatic control system. *Grade 6-7*

1.1 What is homoeostasis?

....

....

....

[2]

1.2 Why are homeostatic control systems important in the body?

....

[1]

1.3 Blood pressure is monitored by sensors in the blood vessels.
Which component of a homeostatic control system senses blood pressure? Tick **one** box.

☐ coordination centre ☐ receptor ☐ stimulus ☐ effector

[1]

1.4 Outline the stages in the negative feedback mechanism when blood pressure becomes too high.

....

....

....

[3]

[Total 7 marks]

2 A person's skin temperature was measured over a 50 minute period. *Grade 6-7*

During that time, the person began exercising. They then returned to a resting state before the end of the investigation. **Figure 1** shows the change in the person's skin temperature over the 50 minutes.

Figure 1

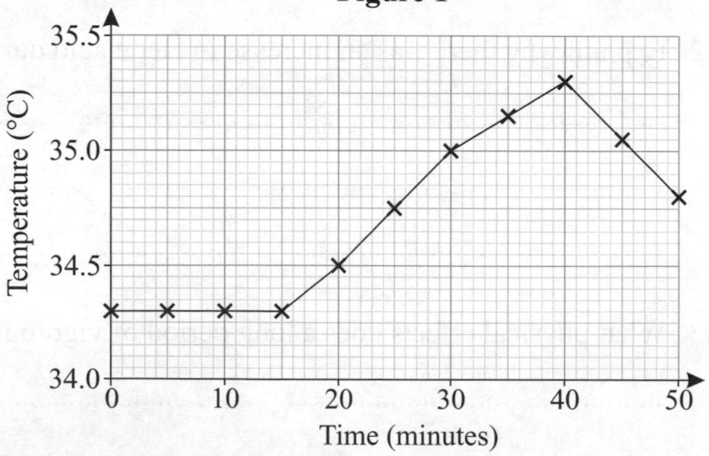

Temperature (°C): 34.0, 34.5, 35.0, 35.5
Time (minutes): 0, 10, 20, 30, 40, 50

2.1 Suggest the time at which the person began exercising.

....

[1]

2.2 Calculate the rate at which the temperature increased between 20 and 30 minutes.

Rate = ... °C/min

[2]

[Total 3 marks]

The Nervous System

1 Figure 1 shows part of the human nervous system. Grade 4-6

1.1 Name the structures labelled **X** and **Y** on **Figure 1**.

X ....

Y ....

[2]

Figure 1

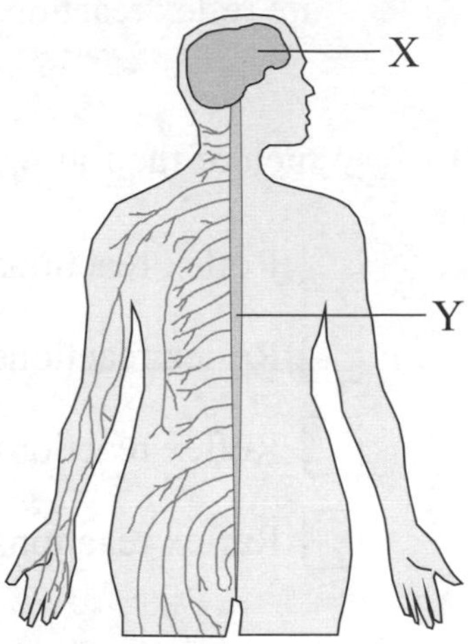

1.2 Which part of the nervous system do structures **X** and **Y** form?

....

[1]

1.3 What is the role of the part of the nervous system formed by structures **X** and **Y**?

....

....

[1]

[Total 4 marks]

2 Multicellular organisms such as humans have a nervous system. Grade 6-7

2.1 What is the function of the nervous system in humans?

....

....

[2]

2.2 Receptor cells in the eye are sensitive to light.
For a nervous system response in the eye, state whether each of the following features is a stimulus, a coordinator or a response.

Spinal cord ....

Bright light ....

Blinking ....

[3]

2.3 Name the **two** main types of neurones found in humans outside the central nervous system.

....

....

[2]

2.4 Name **two** types of effector and state how they respond to nervous impulses.

....

....

[2]

[Total 9 marks]

Synapses and Reflexes

Warm-Up

Circle the examples that are reflex reactions.

Pedalling a bike. The pupil widening in dim light. Dropping a hot plate. Running to catch a bus. Writing a letter.

1 Which of the following sentences is correct? Tick **one** box. Grade 4-6

☐ Reflex reactions are slow and under conscious control.

☐ Reflex reactions are slow and automatic.

☐ Reflex reactions are rapid and automatic.

☐ Reflex reactions are rapid and under conscious control.

[Total 1 mark]

2 **Figure 1** shows a reflex arc. Grade 4-6

Figure 1

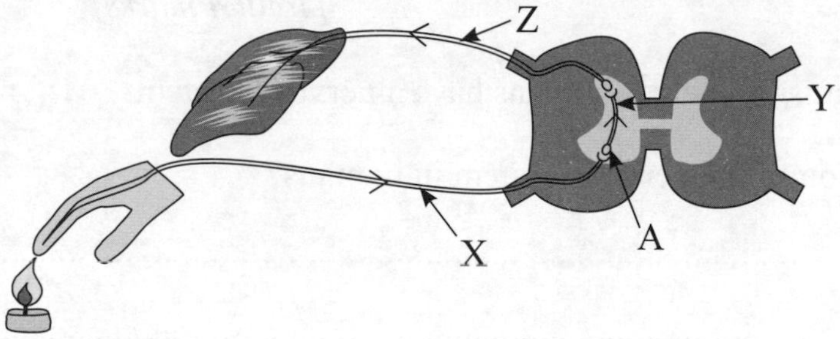

2.1 Name structures **X**, **Y** and **Z**.

X ....

Y ....

Z ....

[3]

2.2 In the reflex arc shown in **Figure 1**, name:

the stimulus ....

the coordinator ....

the effector ....

[3]

2.3 Structure **A** is the junction between two neurones. Name structure **A**.

....

[1]

2.4 How is a nerve signal transmitted across this junction?

....

[1]

[Total 8 marks]

Investigating Reaction Time

PRACTICAL

1 Stimulants, such as caffeine, increase the rate at which nerve impulses travel. An investigation was carried out to assess the impact of different caffeinated drinks on reaction time.

The investigation involved measuring reaction time using a ruler drop test. In this test, a ruler is held above a student's outstretched hand by another person. The ruler is then dropped without warning and the student catches the ruler as quickly as possible. The distance down the ruler where the student caught it is used to calculate their reaction time in seconds (s).
Three different students (Students **1** to **3**) consumed a different caffeinated drink — each one contained a different amount of caffeine. Each student then undertook three ruler drop tests. The results are shown in the table below.

1.1 Calculate the mean reaction time for Student **2** and Student **3**.

	Reaction time (s)		
	Student 1	Student 2	Student 3
Test 1	0.09	0.16	0.20
Test 2	0.10	0.13	0.22
Test 3	0.43	0.15	0.19
Mean	0.21		

Student **2** = ... s

Student **3** = ... s

[2]

1.2 Identify the anomalous result in the table.

....

[1]

1.3 The students' reaction time without any caffeine was **not** measured. Explain why it should have been included in the investigation to assess the effect of each caffeinated drink.

....

....

....

[2]

1.4 Explain why the results of this investigation can't be used to **compare** the effect of the three different caffeinated drinks on reaction time.

....

....

[2]

1.5 An alternative version of the investigation was carried out. This time, the effect of a set quantity of caffeine on the reaction times of different individuals was investigated. Reaction times of three different students were measured, both before and after the consumption of caffeine.
Give **three** variables that should have been controlled in this investigation.

....

....

....

[3]

[Total 10 marks]

The Brain

1 The brain has different regions that carry out different functions. Grade 4-6

1.1 Name the type of cell that makes up most of the brain's material.

....

[1]

1.2 Name the region of the brain that controls unconscious activities.

....

[1]

1.3 State **two** activities that take place in the human body which are not under our conscious control.

....

....

[2]

[Total 4 marks]

2 **Figure 1** shows the human brain. Labels **A, B** and **C** point to three different regions. Grade 6-7

Figure 1

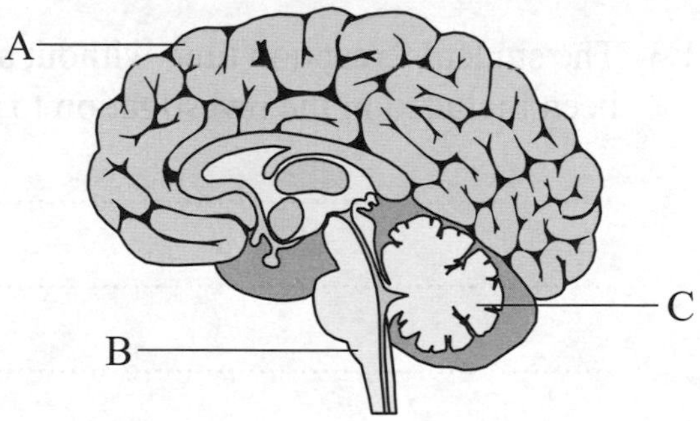

2.1 A hospital patient has had damage to her brain.
She is having problems understanding words.
Which region of the brain is damaged?
Tick **one** box.

☐ A ☐ B ☐ C

[1]

2.2 Which region of the brain controls the heartbeat?
Tick **one** box.

☐ A ☐ B ☐ C

[1]

2.3 Suggest **two** reasons why it is difficult to investigate and treat brain disorders.

....

....

[2]

Scientists have been able to map regions of the brain to their functions using MRI scanning.

2.4 State **two** other ways scientists can gather information to map the regions of the brain to particular functions.

....

....

[2]

[Total 6 marks]

The Eye

Warm-Up

Use the words below to correctly label the diagram of the eye.
You don't have to use every word, but each word can only be used once.

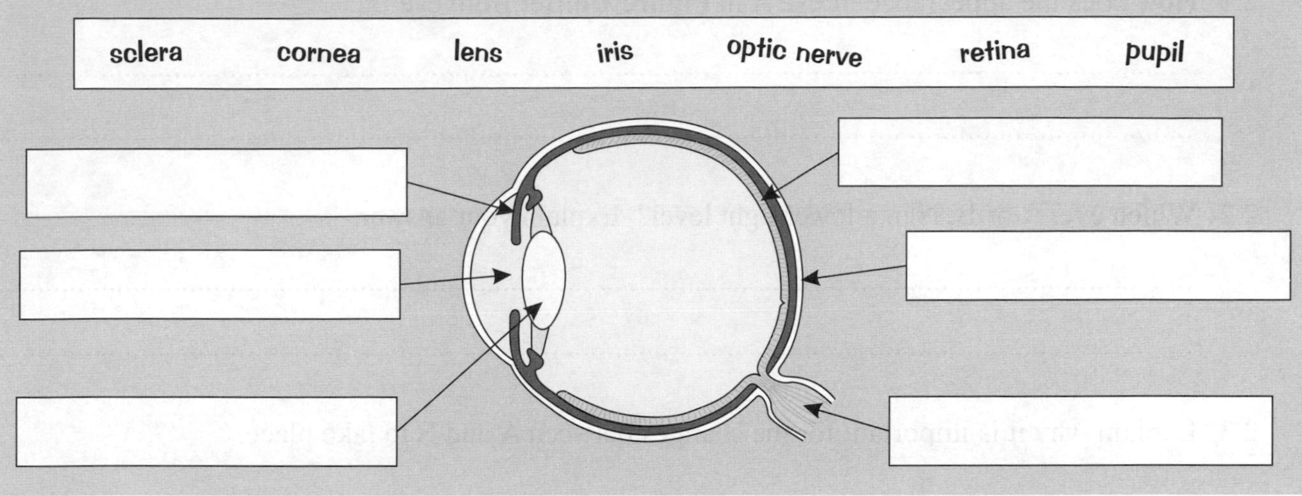

1 The eye is a sense organ containing receptors. Grade 4-6

1.1 Which part of the eye contains receptor cells that are sensitive to light?

.... *[1]*

1.2 The optic nerve carries impulses.
Where do the impulses go once they have left the retina? Tick **one** box.

- ☐ cornea
- ☐ brain
- ☐ sclera
- ☐ lens

[1]

1.3 Name the transparent outer layer at the front of the eye.

.... *[1]*

1.4 Which part of the eye controls the size of the pupil?

.... *[1]*

1.5 Which **two** parts of the eye change the shape of the lens during focussing?

....

.... *[2]*

[Total 6 marks]

2 **Figure 1** shows a human eye as it would appear in two different light levels.

Figure 1

A

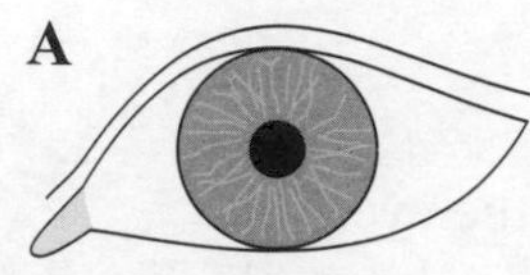

B

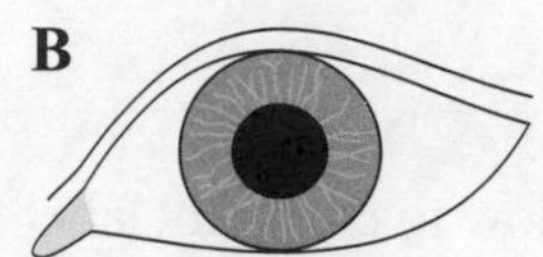

2.1 How does the appearance of eye A in **Figure 1** differ from eye B?

....

....
[2]

2.2 Which eye, A or B, is in a lower light level? Explain your answer.

....

....
[1]

2.3 Explain why it is important for the changes between A and B to take place.

....

....
[2]

[Total 5 marks]

3* Describe how the process of accommodation in the human eye works to:
- focus on a near object,
- focus on a distant object.

Include details of the structures of the eye involved and their functions.

Grade 7-9

....

....

....

....

....

....

....

....

....

....

....

....

[Total 6 marks]

Correcting Vision Defects

1 Some people have to wear spectacle lenses to correct a defect in their vision. **Figure 1** shows how spectacle lenses can be used to correct vision defects.

Figure 1

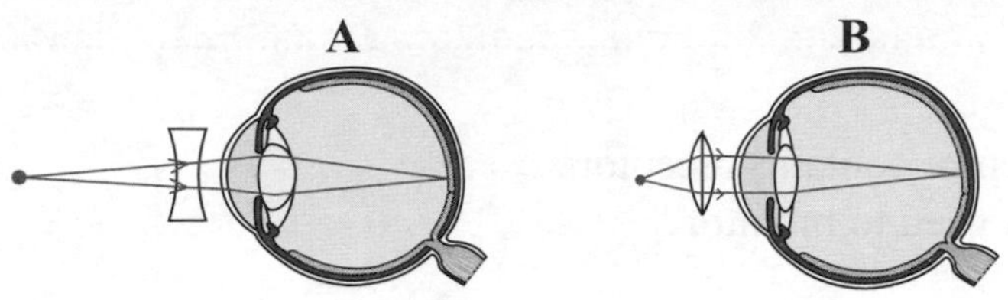

1.1 What vision defect is person **A** in **Figure 1** suffering from?

....

[1]

1.2 Explain how a spectacle lens corrects the sight of a person with this vision defect.

....

....

[2]

1.3 Describe where the rays of light focus in the uncorrected eye of person B in **Figure 1**.

....

[1]

1.4 Name the shape of the spectacle lens used to correct the vision defect of person B.

....

[1]

[Total 5 marks]

2 New technologies are now available to correct vision defects.

Grade 6-7

2.1 Explain how laser technology can be used to correct vision defects.

....

....

[2]

2.2 State **one** other new technology that offers a permanent correction for vision defects.

....

[1]

2.3 Suggest **one** risk of the procedure named in part 2.2.

....

[1]

2.4 A person is suffering from long-sightedness but does not want to wear spectacles or undergo a permanent procedure. What might an optician recommend for this person?

....

[1]

[Total 5 marks]

Controlling Body Temperature

1 The body's thermoregulatory centre monitors and controls body temperature. Grade 4-6

1.1 Name the organ of the body where the thermoregulatory centre is located.

....

[1]

1.2 The thermoregulatory centre contains receptors.
What are these receptors used to monitor?

....

[1]

1.3 Briefly outline how the thermoregulatory centre receives information about the temperature outside the body.

....

....

[2]

[Total 4 marks]

2 The body has several responses to regulate body temperature. Grade 6-7

2.1 Explain how these responses bring about a change in body temperature if the temperature becomes too high.

....

....

....

....

[3]

2.2 Explain how these responses bring about a change in body temperature if the temperature becomes too low.

....

....

....

....

[3]

[Total 6 marks]

Exam Practice Tip

In the exam, you might be asked to explain how the body's temperature regulation mechanisms would respond in a certain situation, e.g. during a cycle race. For these questions, think about what's happening to the body's temperature — if it's going up (e.g. because increased respiration is warming the body), the body's going to want to cool down, and vice versa.

The Endocrine System

1 The endocrine system is a collection of glands in the body that secrete hormones. Grade 4-6

1.1 Which of the following statements about glands is correct? Tick **one** box.

- [ ] Glands secrete hormones directly into cells.
- [ ] Glands secrete hormones directly into tissues.
- [ ] Glands secrete hormones directly into the blood.
- [ ] Glands secrete hormones directly into organs.

[1]

1.2 Which of the following statements best describes hormones?
Tick **one** box.

- [ ] Hormones are tissues.
- [ ] Hormones are chemical molecules.
- [ ] Hormones are cells.
- [ ] Hormones are enzymes.

[1]

1.3 State **two** ways in which the effects of the endocrine system differ from the nervous system.

....

....

[2]

[Total 4 marks]

2 **Figure 1** shows the positions of some glands in the human body. Grade 4-6

2.1 Name glands A to E in **Figure 1**.

A ....

B ....

C ....

D ....

E ....

[5]

Figure 1

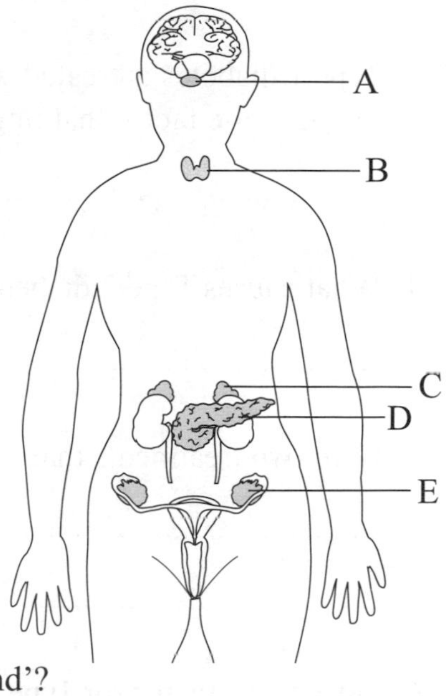

The 'master gland' secretes several hormones into the blood in response to body conditions.

2.2 What is the name of the 'master gland'?

....

[1]

2.3 What is the function of the hormones released by the 'master gland'?

....

....

[2]

[Total 8 marks]

Controlling Blood Glucose

1 The concentration of glucose in the blood is controlled by hormones. (Grade 4-6)

1.1 Which gland in the human body monitors and controls blood glucose concentration?
Tick **one** box.

☐ pancreas ☐ pituitary gland ☐ thyroid ☐ testis

[1]

1.2 Which hormone is produced when blood glucose concentration becomes too high?

....

[1]

1.3 Describe what happens to excess glucose in the blood.

....

....

[2]

[Total 4 marks]

2 Diabetes exists in two different forms, Type 1 and Type 2. (Grade 6-7)

2.1 What causes Type 1 diabetes?

....

[1]

2.2 What is the defining characteristic of Type 1 diabetes?

....

[1]

2.3 Type 1 diabetes is treated with insulin injections.
Suggest **one** factor that might affect the amount of insulin injected by a patient.

....

[1]

2.4 What causes Type 2 diabetes?

....

[1]

2.5 Give **two** treatments that a doctor would recommend for Type 2 diabetes.

....

....

[2]

2.6 Give a risk factor for Type 2 diabetes.

....

[1]

[Total 7 marks]

3 In an experiment, the blood glucose concentration of a person without diabetes was recorded at regular intervals in a 90 minute time period. Fifteen minutes into the experiment, a glucose drink was given. **Figure 1** shows the results of the experiment.

Figure 1

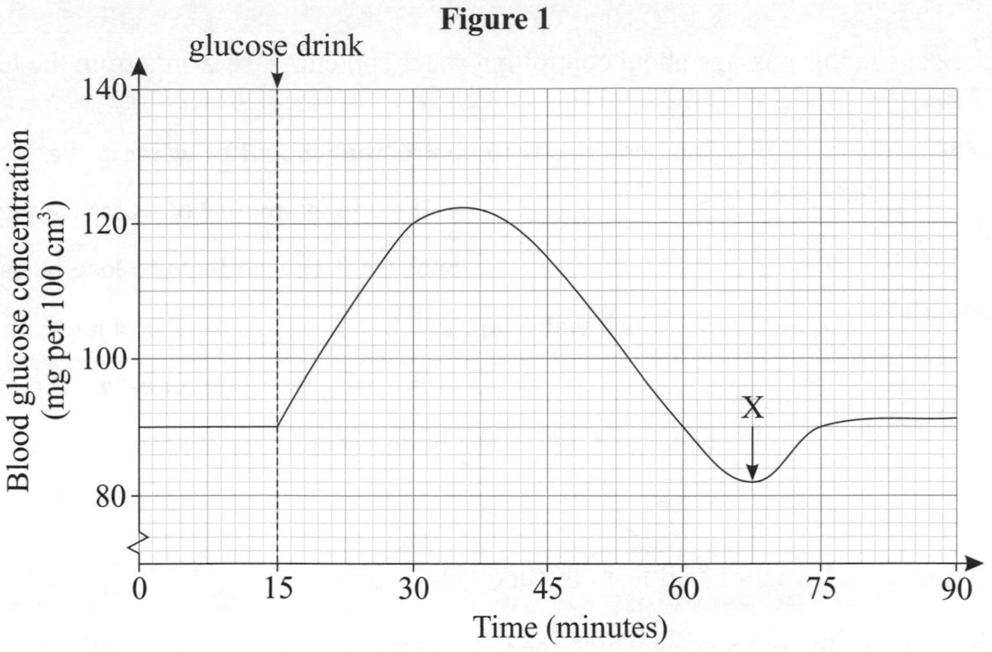

3.1 Explain what is happening to the blood glucose concentration between 15 and 60 minutes.

....

....

....

[3]

3.2 Name the hormone being released by the pancreas at point **X** on the graph.

....

[1]

3.3 Describe the effect that hormone **X** has on the blood glucose concentration.

....

[1]

3.4 Explain how hormone **X** causes this effect.

....

....

[1]

3.5 Suggest how the shape of the graph would differ if the person had Type 1 diabetes.

....

....

[1]

[Total 7 marks]

Exam Practice Tip

There are a few similar-sounding names when it comes to the control of blood glucose, so make sure you've got your head around which is which (and how to spell them). You won't get the mark if, for example, you write about 'glucogen'...

The Kidneys

Warm-Up

Fill in the gaps in this passage about controlling water content. Use words from the left.

pancreas
kidneys
blood
enzymes
osmosis
cells
ions
active transport

The body needs to control the amount of water in the .... The wrong amount of water can damage .... because it causes them to lose or gain too much water by .... The amount of .... in the body also affects cells. The amount of water and ions in the body is controlled by the ....

1 Unwanted substances are removed from the body in the urine. Grade 4-6

1.1 What is the name of the process by which the kidneys produce urine? Tick **one** box.

- [ ] active transport
- [ ] filtration
- [ ] osmosis
- [ ] diffusion

[1]

1.2 The unwanted substances removed in the urine include excess water and some ions. Which other substance is also removed via the kidneys in the urine?

....

[1]

A process in the kidneys returns useful substances to the blood so that they are not lost in urine.

1.3 What is the name of this process?

....

[1]

1.4 Name **two** useful substances that are returned to the blood by this process.

....

....

[2]

[Total 5 marks]

2 Some water leaves the body in the urine. Water can also be lost from the skin by sweating. Grade 4-6

2.1 Give **one** other way that water can be lost from the body.

....

[1]

2.2 Name **two** substances other than water that are lost through the skin in sweat.

....

....

[2]

2.3 Which of the following statements is true?
Tick **one** box.

☐ The body can control water loss from the skin.

☐ The body can only control water loss from the skin at night.

☐ The body can't control water loss from the skin.

[1]

[Total 4 marks]

3 Excess amino acids in the body are broken down. Grade 4-6

3.1 In which organ of the body are the excess amino acids broken down?

....

[1]

3.2 What is the source of these excess amino acids?

....

[1]

3.3 Deamination forms part of the breakdown process. Name the waste product of deamination.

....

[1]

3.4 Explain why this waste product is converted into urea for excretion as soon as it is formed.

....

[1]

[Total 4 marks]

4* The body is constantly monitoring and regulating its water content. Describe the body's response when the brain detects that the water content is too low. Include details of any hormones and structures involved. Grade 7-9

....

....

....

....

....

....

....

....

[Total 4 marks]

Kidney Failure

1 **Figure 1** shows the operation of a dialysis machine. 

Figure 1

partially permeable membrane
dialysis fluid out
waste substances diffuse out
dialysis fluid in
blood from body
blood back to body

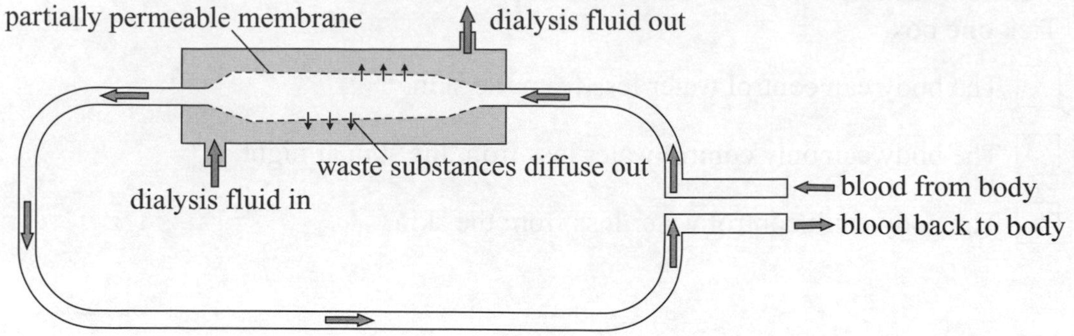

1.1 Why might a person need to use a dialysis machine?

....

....

[1]

In a dialysis machine, the person's blood flows alongside a partially permeable membrane surrounded by dialysis fluid.

1.2 Name **one** waste substance that will move through the partially permeable membrane.

....

[1]

1.3 Suggest why proteins cannot pass through the partially permeable membrane.

....

[1]

1.4 Suggest why it is important that the dialysis fluid has the same concentration of dissolved substances as healthy blood.

....

[1]

1.5 Name **two** substances that should be present in the dialysis fluid.

....

....

[2]

[Total 6 marks]

2 Suggest **one** advantage and **one** disadvantage of treating organ failure, such as kidney failure, by an organ transplant rather than a mechanical device. Grade 7-9

Advantage: ....

....

Disadvantage: ....

....

[Total 2 marks]

Puberty and the Menstrual Cycle

1 The release of sex hormones begins at puberty. Grade 4-6

1.1 What is the name of the main female hormone produced in the ovary? Tick **one** box.

☐ progesterone ☐ oestrogen ☐ luteinising hormone ☐ follicle stimulating hormone

[1]

1.2 What is the name of the process by which eggs are released from the ovary?

....

[1]

1.3 How often is an egg released from an ovary? Tick **one** box.

☐ Every 7 days. ☐ Every 14 days. ☐ Every 21 days. ☐ Every 28 days.

[1]

1.4 Name the hormone that stimulates the release of an egg.

....

[1]

1.5 Name the hormone that stimulates sperm production in men.

....

[1]

1.6 Where in the male body is this hormone produced?

....

[1]

[Total 6 marks]

2 Four main hormones interact with each other in the control of the menstrual cycle. Grade 6-7

2.1 Which two hormones are involved in maintaining the uterus lining?

....

....

[2]

2.2 What is the name of the gland that secretes follicle stimulating hormone (FSH)?

....

[1]

2.3 State **two** effects of FSH during the menstrual cycle of a woman.

....

....

[2]

2.4 Which hormone stimulates the release of luteinising hormone (LH)?

....

[1]

[Total 6 marks]

Controlling Fertility

Warm-Up

Sort the methods of contraception into the correct places in the table.

abstinence
contraceptive injection
condom
diaphragm
plastic intrauterine device
sterilisation
contraceptive patch

Hormonal	Non-hormonal

1 Some methods of contraception use hormones to control the fertility of a woman. Grade 4-6

1.1 How is an oral contraceptive taken into the body?
Tick **one** box.

☐ As an injection.

☐ As a tablet taken by mouth.

☐ Through the skin from a patch.

[1]

1.2 How do oral contraceptives containing multiple hormones prevent pregnancy?
Tick **one** box.

☐ The hormones inhibit oestrogen production.

☐ The hormones inhibit FSH production.

☐ The hormones inhibit LH production.

[1]

1.3 The contraceptive implant is inserted under the skin of the arm.
Which hormone does it release?

....

[1]

1.4 How does the hormone released by the contraceptive implant prevent pregnancy?

....

[1]

[Total 4 marks]

2 Fertility can be controlled by non-hormonal methods of contraception. Grade 4-6

2.1 Name a barrier method of contraception that can be used by men.

....
[1]

2.2 Name a barrier method of contraception that can be used by women.

....
[1]

2.3 How do barrier methods of contraception prevent a woman becoming pregnant?

....
[1]

2.4 What is the name given to chemicals that kill or disable sperm?

....
[1]

2.5 A couple not wishing to have children do not want to use any form of contraception. Suggest how they could avoid pregnancy.

....
[1]

2.6 Name a surgical method of controlling fertility that can be carried out in both men and women.

....
[1]

2.7 Name a barrier method of contraception that protects against sexually transmitted infections.

....
[1]

[Total 7 marks]

3 A woman is considering which contraceptive to use. Grade 7-9

3.1 Suggest **one** advantage of choosing the contraceptive injection over the contraceptive pill.

....
[1]

3.2 Suggest **one** disadvantage of choosing the contraceptive injection over the contraceptive pill.

....
[1]

3.3 Suggest **one** advantage of choosing a barrier method of contraception over a hormonal contraceptive.

....
[1]

[Total 3 marks]

Exam Practice Tip

Knowing how the hormones that control the menstrual cycle interact with each other can be handy when it comes to understanding how these hormones are used to control fertility. So make sure you've got it all sorted out in your head.

More on Controlling Fertility

1 A couple want to have children but the woman has not yet become pregnant. Blood tests have shown that she has a low level of follicle stimulating hormone (FSH). She is treated with a fertility drug.

Grade 6-7

1.1 Explain why a low level of FSH may be preventing the woman from becoming pregnant.

....

[1]

1.2 In addition to FSH, which other hormone will the fertility drug contain to help the woman become pregnant? Give a reason for your answer.

....

....

[2]

1.3 Suggest **one** advantage and **one** disadvantage of this method of fertility treatment.

Advantage: ....

Disadvantage: ....

[2]

[Total 5 marks]

2 *In vitro* fertilisation is a reproductive treatment that can help people with fertility problems have children.

Grade 7-9

2.1 Describe the stages involved in a course of *in vitro* fertilisation treatment.

....

....

....

....

....

....

....

....

[5]

2.2 Give **two** disadvantages of *in vitro* fertilisation treatment.

....

....

....

[2]

[Total 7 marks]

Calum

Adrenaline and Thyroxine

Warm-Up

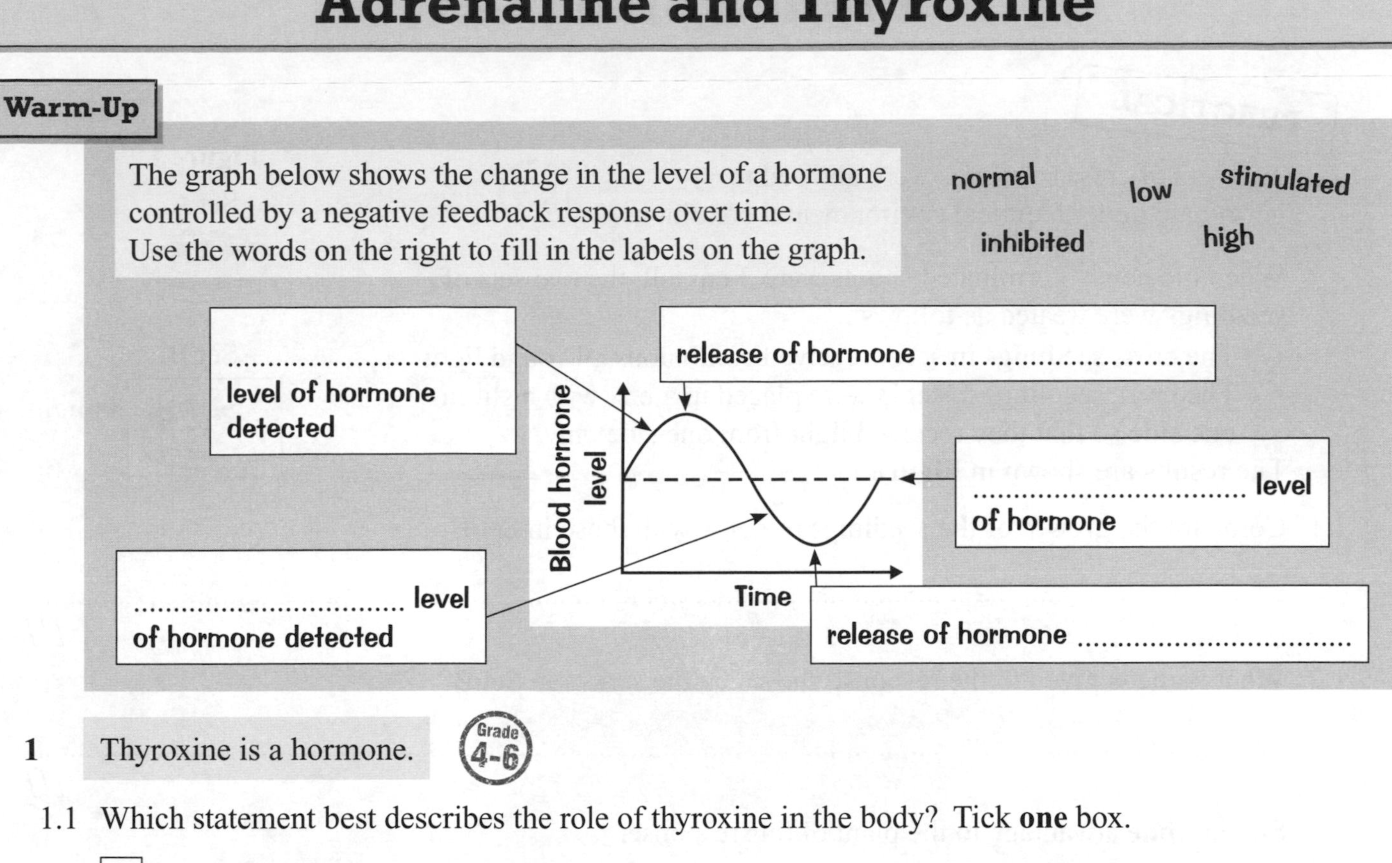

1 Thyroxine is a hormone. Grade 4-6

1.1 Which statement best describes the role of thyroxine in the body? Tick **one** box.

- [ ] Thyroxine inhibits development.
- [ ] Thyroxine regulates basal metabolic rate.
- [ ] Thyroxine decreases heart rate.

[1]

1.2 Which gland produces thyroxine?

....

[1]

[Total 2 marks]

2 The hormone adrenaline is produced in times of fear or stress. Grade 6-7

2.1 Where is adrenaline released from?

....

[1]

2.2 Describe the effect that adrenaline has on the body.

....

....

....

[3]

2.3 Name the response that adrenaline prepares the body for.

....

[1]

[Total 5 marks]

Plant Hormones

PRACTICAL

1 Two sets of cress seedlings were allowed to germinate under identical environmental conditions. 

Figure 1

Set A

Set B

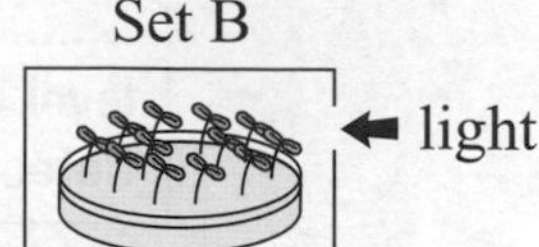

When the newly germinated shoots were 3 cm tall, the two sets of seedlings were treated as follows:

- The cress seedlings in set A received continuous all-round light.
- The cress seedlings in set B were placed in a box with a slit in one side so that they received light from one side only.

The results are shown in **Figure 1**.

1.1 Compare the growth of the seedlings in Set A with those in Set B.

....

[1]

1.2 What name is given to the response shown by the shoots in Set B?

....

[1]

1.3 Suggest **one** advantage to the plant of this response.

....

[1]

Auxin is a hormone that controls the growth of a plant in response to light.

1.4 Explain the results for Set B. Refer to auxin in your answer.

....

....

....

[3]

[Total 6 marks]

2 A tropism is growth by plants in response to a stimulus. Tropisms are positive when the plant, or part of the plant, grows towards a stimulus. They are negative when the plant, or part of the plant, grows away from a stimulus.

Grade 7-9

2.1 What type of tropism is shown by a root growing towards gravity?

....

[1]

2.2 Name a part of a plant that shows a negative tropism in response to gravity.

....

[1]

2.3 In which direction does a root showing negative phototropism grow?

....

[1]

[Total 3 marks]

Commercial Uses of Plant Hormones

1 Plants produce hormones to coordinate and control a variety of processes. Grade 4-6

1.1 Which hormone controls cell division in plants?

.... *[1]*

1.2 Which of the following processes do gibberellins initiate in plants? Tick **one** box.

☐ root growth ☐ seed germination ☐ fruit ripening

[1]

[Total 2 marks]

2 Give **three** uses of gibberellins in horticulture. Grade 4-6

....

....

....

[Total 3 marks]

3 The bananas in UK supermarkets often come from countries abroad, such as Ecuador. Grade 6-7

3.1 Name the hormone commonly used in the food industry to control the ripening of fruit.

.... *[1]*

3.2 Suggest why bananas destined for UK supermarkets are picked before they are ripe.

....

.... *[2]*

[Total 3 marks]

4 A gardener wants to clone a plant in his garden. He takes a cutting of the plant and dips it into a powder containing a particular hormone. Grade 6-7

4.1 Suggest which hormone the powder contains.

.... *[1]*

4.2 Suggest what the powder is for.

.... *[1]*

4.3 What other product might the gardener use in his garden that may contain this hormone?

.... *[1]*

[Total 3 marks]

DNA

1 DNA makes up the genetic material in animal and plant cells. Grade 4-6

1.1 Which of the following statements about DNA is correct?
Tick **one** box.

☐ DNA is located in the cytoplasm of animal and plant cells.

☐ DNA is located in the ribosomes in animal and plant cells.

☐ DNA is located in the nucleus of animal and plant cells.

☐ DNA is located in vacuoles in animal and plant cells.

[1]

1.2 What are chromosomes?
Tick **one** box.

☐ Proteins coded for by DNA.

☐ The structures that contain DNA.

☐ The site of protein synthesis.

☐ The bases that make up DNA.

[1]

[Total 2 marks]

2 An organism's DNA contains lots of sections called genes. Grade 6-7

2.1 Outline the function of genes.

....

....

[2]

2.2 What is meant by the term 'genome'?

....

....

[1]

2.3 Give **one** reason why it is important for scientists to have an understanding of the human genome.
Explain your answer.

....

....

....

....

....

[2]

[Total 5 marks]

The Structure of DNA and Protein Synthesis

1 DNA is made up of long chains of nucleotides.
Each nucleotide contains one of four DNA bases. (Grade 6-7)

1.1 What are the four bases found in DNA?
Tick **one** box.

☐ A, T, P and G ☐ C, T, G and F ☐ A, C, G and T ☐ T, C, A and E

[1]

Figure 1 shows a DNA nucleotide.

Figure 1

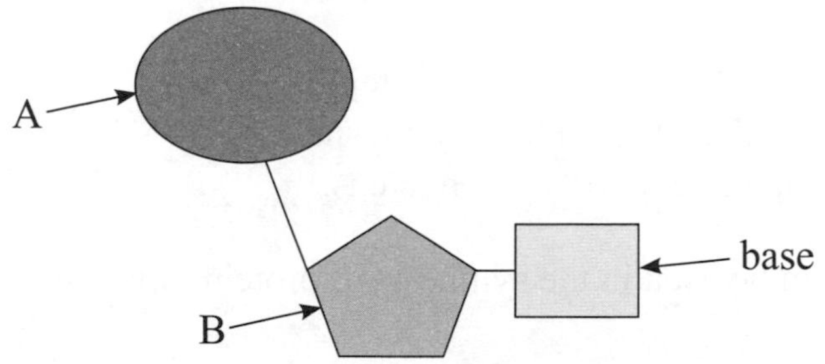

A
B
base

1.2 Name the parts labelled A and B.

A: .... B: ....

[2]

Bases in the two strands that make up a DNA molecule always pair up in the same way.
Figure 2 shows a short piece of DNA.

Figure 2

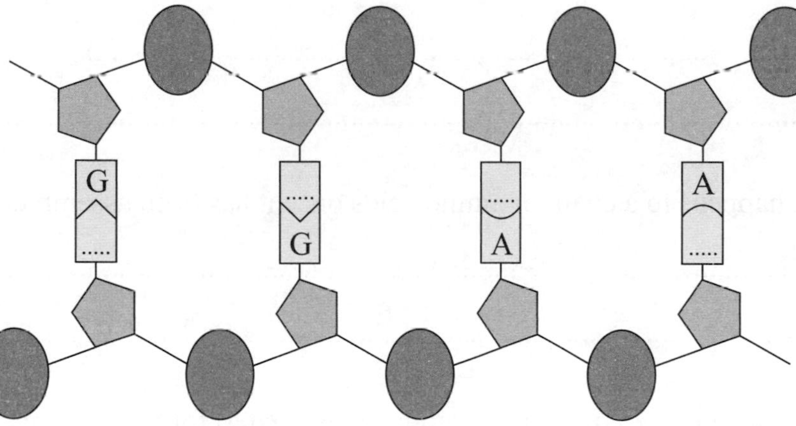

G ..... | ..... G | ..... A | A .....

1.3 Complete **Figure 2** by writing the correct letter in the unlabelled bases.

[2]

1.4 Explain how the DNA bases in a gene code for a specific chain of amino acids.

....

....

....

[2]

[Total 7 marks]

2 DNA molecules carry the code to make proteins. (Grade 6-7)

2.1 Name the structures in a cell where protein synthesis takes place.

....

[1]

2.2 Not all parts of DNA code for amino acids. Some of the DNA is non-coding.
Briefly describe **one** role that non-coding DNA plays in protein synthesis.

....

[1]

[Total 2 marks]

3 DNA is located in the nucleus of the cell. Protein synthesis takes place outside of the nucleus in the cytoplasm. DNA molecules are too large to leave the nucleus. (Grade 6-7)

3.1 Explain how DNA can be used in the synthesis of proteins when it is unable to leave the nucleus.

....

....

....

[2]

3.2 Describe the process that takes place at the site of protein synthesis to produce chains of amino acids.

....

....

....

[3]

3.3 Explain what happens to a chain of amino acids once it has been assembled.

....

....

[2]

3.4 Proteins have many jobs. For example, collagen is used to form structures in the body.
Briefly outline the role of **two** other types of protein found in the human body.

....

....

....

....

[4]

[Total 11 marks]

Mutations

Warm-Up

Draw circles to show whether the statements below are **true** or **false**.

All mutations affect the function of the protein coded for.	True	/	False
Mutations occur continuously.	True	/	False
Mutations cannot be inherited.	True	/	False

1 Mutations are random changes in the DNA of an organism. 

Figure 1 shows the normal order of bases in a section of DNA and then the same section of DNA after a mutation has occurred.

Figure 1

DNA before mutation: A A G C T T C C A

DNA after mutation: A A G C T T C C G A

1.1 Circle where the mutation has taken place in **Figure 1**.

[1]

1.2 Explain how mutations could lead to a change in the protein being synthesised.

....

....

....

....

[3]

Mutations can affect the functioning of a protein.

1.3 Suggest **one** possible effect of a mutation that changes the shape of a structural protein.

....

....

....

[2]

1.4 Suggest **one** possible effect of a mutation in a gene that codes for a particular enzyme.

....

....

....

[3]

[Total 9 marks]

Reproduction

1 Reproduction can be sexual or asexual. Grade 4-6

In sexual reproduction, gametes from a male and female fuse together.

1.1 Name the male gamete in animals.

....

[1]

1.2 Name the female gamete in plants.

....

[1]

1.3 Which type of cell division is involved in the production of gametes?

....

[1]

Asexual reproduction does not involve gametes.

1.4 What name can be given to the cells resulting from asexual reproduction?

☐ gametes ☐ clones ☐ eggs ☐ chromosomes

[1]

1.5 Name the type of cell division used in asexual reproduction.

....

[1]

[Total 5 marks]

2 Sexual reproduction involves the fusion of gametes to form a fertilised egg cell. Grade 6-7

2.1 Explain how a fertilised egg cell has the correct number of chromosomes.

....

....

....

[2]

2.2 Asexual and sexual reproduction are very different methods.
Give **four** ways in which asexual reproduction is different to sexual reproduction.

....

....

....

....

....

....

[4]

[Total 6 marks]

Meiosis

1 Sexual reproduction in humans involves meiosis. Grade 4-6

1.1 Where in the body does meiosis take place?

....

[1]

1.2 What happens to the DNA at the very start of meiosis, before the cell starts to divide?

....

[1]

1.3 How many cell divisions are there during the process of meiosis?

....

[1]

1.4 Briefly describe the results of meiosis.

....

....

....

....

[3]

[Total 6 marks]

2 During fertilisation, two gametes formed by meiosis join together. Grade 4-6

2.1 How many copies of each chromosome does the resulting cell have?

....

[1]

After the two gametes join to produce a fertilised egg, the cells divide repeatedly.

2.2 What type of cell division do these cells undergo?

....

[1]

2.3 The dividing cells form an embryo.
What happens to the cells in the embryo as it develops in order to form the whole organism?

....

....

[1]

[Total 3 marks]

Exam Practice Tip

It's outrageously easy to get mixed up between meiosis and mitosis when the pressure is on in an exam. Remember, meiosis is the one that makes — eggs and sperm. Mitosis makes twin (identical) cells. Even if you know the difference, it's still really easy to accidentally write one when you mean the other 'cos the words are so similar, so always check twice.

More on Reproduction

Warm-Up

Complete the passage using words from below. You don't need to use every word.

runners　　different　　spores　　stalks　　identical　　seeds

Some organisms can reproduce both sexually and asexually. For example, strawberry plants reproduce asexually using .... and reproduce sexually by producing ....

Asexual reproduction in strawberry plants results in genetically .... offspring whereas sexual reproduction produces genetically .... offspring.

1 Some organisms can reproduce both sexually and asexually. (Grade 4-6)

1.1 Daffodils can reproduce by producing seeds or bulbs. Bulbs divide to form new plants. What type of reproduction is shown when the bulbs divide to produce new plants?

....

[1]

1.2 Which of these statements about the reproduction of the malaria parasite is true? Tick **one** box.

- [ ] Malaria parasites reproduce sexually in the human host, but asexually in the mosquito.
- [ ] Malaria parasites reproduce sexually in both the human host and the mosquito.
- [ ] Malaria parasites reproduce asexually in the human host, but sexually in the mosquito.
- [ ] Malaria parasites reproduce asexually in both the human host and the mosquito.

[1]

[Total 2 marks]

2 Sexual and asexual reproduction each have their own advantages and disadvantages. (Grade 6-7)

2.1 State **two** advantages of asexual reproduction over sexual reproduction.

....

....

[2]

2.2 An unfavourable environmental change affects a population of organisms. Explain why sexual reproduction increases the chance of the population surviving.

....

....

....

....

....

[4]

[Total 6 marks]

X and Y Chromosomes

1 In humans, the gender of offspring is determined by a pair of sex chromosomes — X and Y.

1.1 Including the sex chromosomes, how many chromosomes are there in a normal body cell? Tick **one** box.

☐ 22 single chromosomes ☐ 22 pairs of chromosomes

☐ 23 pairs of chromosomes ☐ 23 single chromosomes

[1]

Figure 1 shows how the gender of offspring is determined.

Figure 1

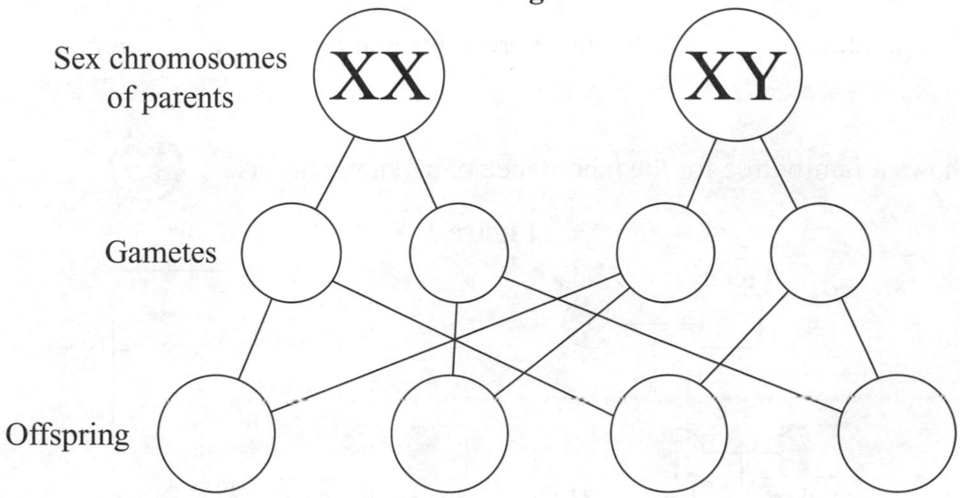

Sex chromosomes of parents — XX — XY

Gametes

Offspring

1.2 Circle the male parent in **Figure 1**.

[1]

1.3 Fill in the sex chromosomes of the gametes produced by each parent in **Figure 1**.

[1]

1.4 Complete **Figure 1** to show the combination of sex chromosomes in the offspring.

[1]

1.5 What is the ratio of male to female offspring?

....

[1]

1.6 Sex determination can also be shown in a Punnett square.
Produce a Punnett square to show how the gender of offspring is determined.

[2]

[Total 7 marks]

Genetic Diagrams

Warm-Up

Use the words and phrases to complete the passage below. You don't have to use every one.

homozygous dominant multiple genes genotypes homologous alleles recessive heterozygous a single gene

Genes exist in different versions called .... These can be dominant or .... If an individual has two copies of the same version of a gene, they are said to be ...., but if they have two different versions, they are said to be .... Some characteristics are controlled by ...., but most are controlled by ....

1 **Figure 1** shows a family tree for the inheritance of a genetic disease.

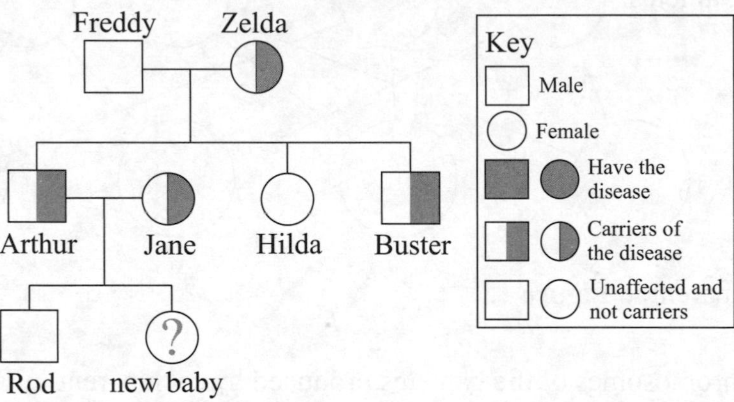

1.1 How can you tell that the allele for the disease is not dominant?

....

....

[1]

1.2 The alleles for the disease are D and d. Both Arthur and Jane are carriers of the disease. Complete the Punnett square in **Figure 2** to determine the probability of their new baby being unaffected and not a carrier of the disease.

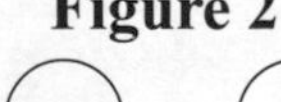

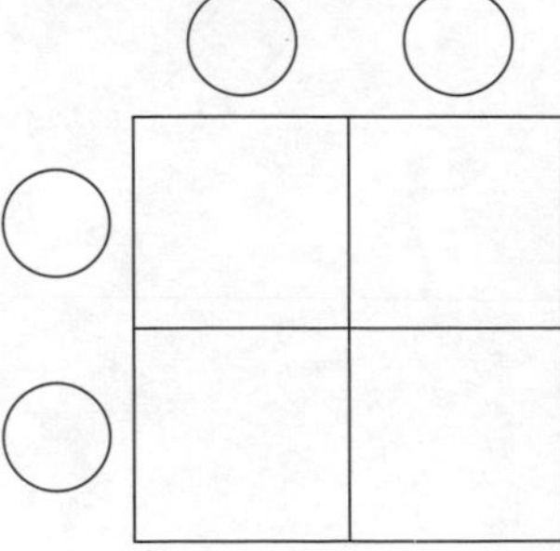

probability = ... %

[4]

[Total 5 marks]

2 Hair length in dogs is controlled by two alleles. Short hair is caused by the allele 'H' and long hair is caused by the allele 'h'.

Figure 3 shows a genetic diagram of a cross between two dogs with the genotype Hh.

Figure 3

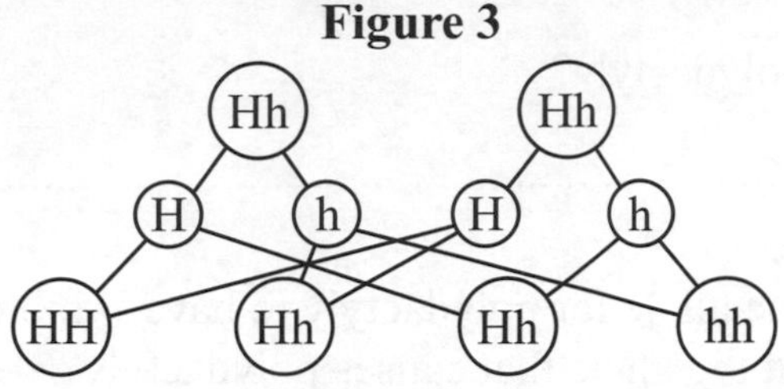

2.1 What is the expected ratio of short-haired puppies to long-haired puppies in this cross?

....

[1]

2.2 A dog with the genotype HH was crossed with a dog with the genotype hh. They had 8 puppies. How many of those puppies would you expect to have short hair?

Complete **Figure 4** below to explain your answer.

Figure 4

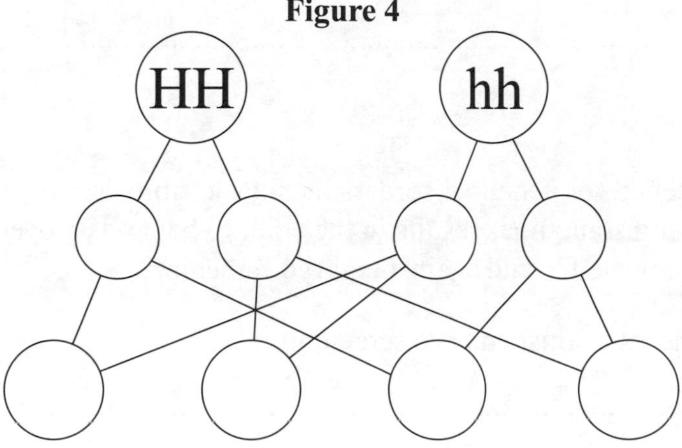

number of short-haired puppies = ...

[3]

2.3 A male dog heterozygous for short hair was then crossed with a female dog homozygous for long hair. What would you expect the ratio of long-haired to short-haired puppies to be in the offspring of this cross?

Construct a Punnett square to explain your answer.

ratio = ...

[3]

[Total 7 marks]

Inherited Disorders

1 Polydactyly and cystic fibrosis are examples of inherited disorders. Grade 6-7

1.1 What are the symptoms of polydactyly?

....

[1]

1.2 A person only has to have one allele for polydactyly to have symptoms.
What does this tell you about the allele that causes polydactyly?

....

[1]

1.3 Even if both parents each carry one copy of the allele that causes cystic fibrosis, there is only a relatively small chance that their offspring will have the disorder. Explain why this is the case.

....

....

....

[3]

[Total 5 marks]

2 Embryos can be screened for genetic disorders like cystic fibrosis.
The results of screening sometimes results in the embryo being destroyed.
There are lots of arguments for and against embryo screening. Grade 7-9

2.1 Suggest **three** arguments against embryo screening.

....

....

....

....

....

....

[3]

2.2 Suggest **three** arguments for embryo screening.

....

....

....

....

....

....

[3]

[Total 6 marks]

The Work of Mendel

Warm-Up

Circle the correct words or phrases below so that the passage is correct.

Gregor Mendel was an Austrian monk who studied mathematics and natural history at the University of Vienna. He is best known for his work on speciation/evolution/genetics. In the late 18th century/early 19th century/mid-19th century, he performed many breeding experiments using fungi/plants/pigs. Mendel noted that offspring often shared characteristics with their parents. He proposed that characteristics were passed on/lost/altered from one generation to the next in units.

1 Scientists didn't realise the importance of Gregor Mendel's research until after his death.

1.1 Suggest why scientists at the time didn't understand how important Mendel's work was.

....

....

....

[1]

1.2 Briefly outline the discoveries made after Mendel's work that built on his discovery of 'hereditary units'. Include how these led to our current understanding of genes.

....

....

....

....

....

....

....

....

....

....

....

[5]

[Total 6 marks]

Variation

1 Variation occurs in many different organisms. 

1.1 Dalmations and pugs are both members of the same species. However, they look very different. For example, dalmations have spots but pugs do not.
What type of variation is causing this difference?

....

[1]

Figure 1 and **Figure 2** show two plants of the same species growing in opposite corners of a garden. The plant in **Figure 1** was grown in a sunny corner, whereas the plant in **Figure 2** was grown in a shady corner.

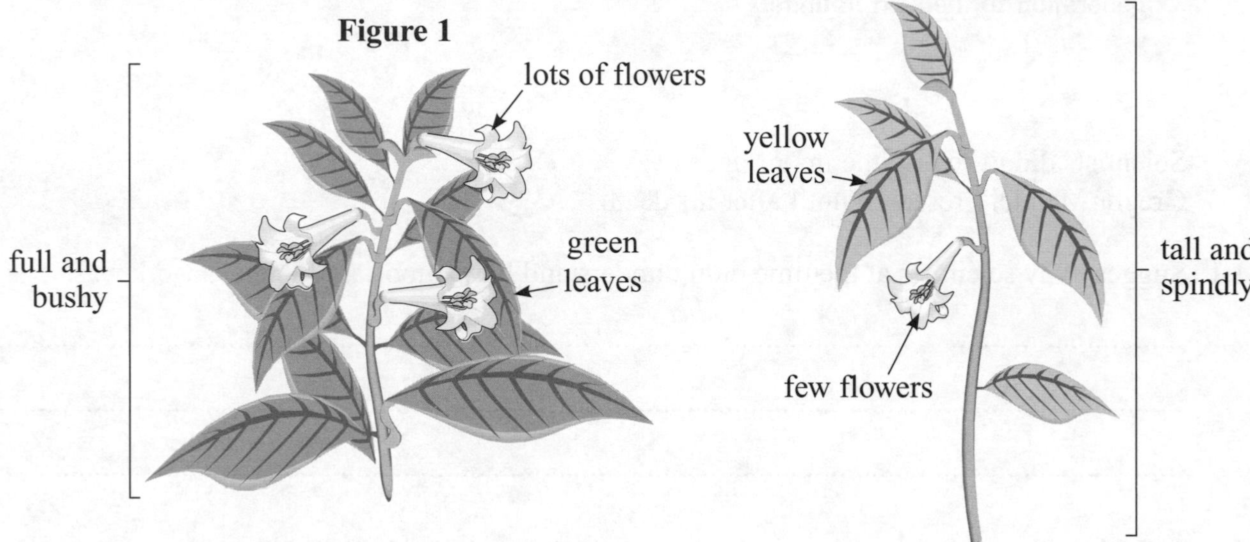

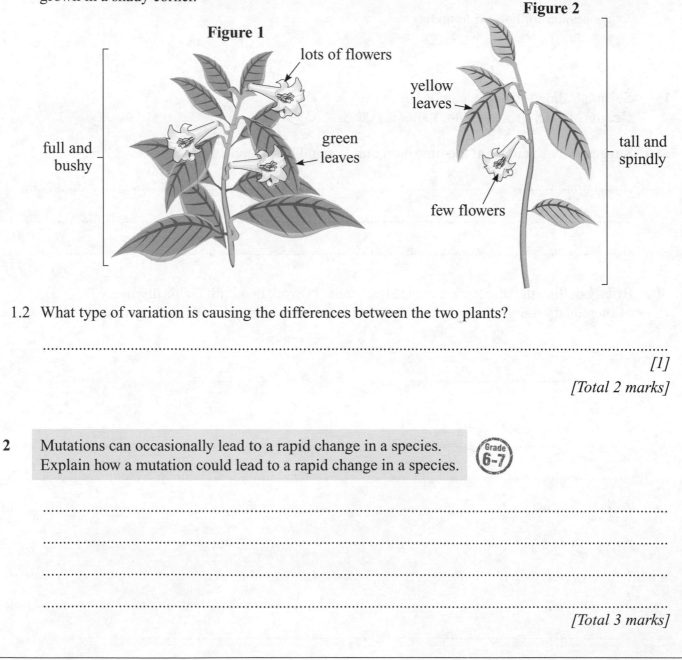

1.2 What type of variation is causing the differences between the two plants?

....

[1]

[Total 2 marks]

2 Mutations can occasionally lead to a rapid change in a species.
Explain how a mutation could lead to a rapid change in a species. 

....

....

....

....

[Total 3 marks]

Exam Practice Tip

Remember that while variation can be caused by either genetic or environmental factors, it's usually caused by a mixture of both interacting with each other. In the exam, you might get an example of variation that you've never heard of before. Don't worry if you do, all the information you need to answer the question will be there. Just apply your knowledge.

Evolution

1 Evolution by natural selection can sometimes result in the formation of two new species with very different phenotypes. Grade 4-6

1.1 What is the name of the process by which new species form?

....

[1]

1.2 Explain how you could know for certain if two populations of one original species had become two new species.

....

....

[1]

[Total 2 marks]

2 The theory of evolution is that all species of living things have evolved from simple life forms that first developed many years ago. Grade 4-6

2.1 How many years ago did the simple life forms develop?
Tick **one** box.

☐ Less than 3 million years ago

☐ More than 3 billion years ago

☐ Less than 300 thousand years ago

☐ More than 5 billion years ago

[1]

2.2 Charles Darwin is best known for proposing the theory of evolution by natural selection.
Some of the evidence that he based his theory on was gathered on his round-the-world trip.
Give **two** other sources of developing knowledge used by Darwin.

....

[2]

When Darwin proposed his theory, he wasn't able to fully explain why everything happened.
Scientists have since been able to use new knowledge of genetics to develop the theory.

2.3 Darwin wasn't able to explain how new variations in phenotype occurred.
What explanation has since been found for new variations in phenotype?

....

[1]

2.4 Darwin was also unable to explain how characteristics are passed on to offspring.
What explanation has since been found for how characteristics are passed on?

....

....

[1]

[Total 5 marks]

3 Extinction is when a species completely dies out because they're not able to evolve quickly enough to adapt to a change in their environment. Give **five** factors that can cause a species to become extinct.

Grade 6-7

[Total 5 marks]

4 **Figure 1** and **Figure 2** show two hares. The hare in **Figure 1** lives in a very cold climate. The hare in **Figure 2** lives in a warmer climate.

Grade 7-9

Figure 1

Figure 2

The hare in **Figure 2** uses its large ears as a cooling mechanism. They allow lots of heat to leave the hare's body. The hare in **Figure 1** has smaller ears.

Suggest how the species of hare in **Figure 1** evolved to have smaller ears than hares that live in warmer climates.

[Total 5 marks]

More About Evolution

1 When Darwin proposed his theory in 1859, it was highly controversial. Grade 6-7

1.1 What was the title of the book in which Charles Darwin proposed his theory of evolution by natural selection?
Tick **one** box.

- [ ] On the Theory of Evolution
- [ ] On the Origin of Species
- [ ] On the Process of Natural Selection
- [ ] On the Progression of Organisms

[1]

1.2 Darwin did not have an explanation for how characteristics were passed on.
Give **two** other reasons why Darwin's theory was controversial at the time it was published.

....

....

....

....

....

[2]

Other theories on evolution existed at the same time as Darwin's.
One of these was proposed by Jean-Baptise Lamarck.

1.3 What did Lamarck believe about the mechanism of evolution?

....

....

....

[1]

1.4 Lamarck's theory was rejected because experiments didn't support his hypothesis.

Darwin's theory is now widely accepted because lots of evidence supports it, such as thc discovery of how characteristics are passed on by genes.

Give **two** other examples of evidence that supports the theory of evolution by natural selection.

....

....

....

....

[2]

[Total 6 marks]

Selective Breeding

1 Selective breeding is used in several different industries. (Grade 4-6)

1.1 Which of these is another name for the process of selective breeding?
Tick **one** box.

☐ Evolution ☐ Natural selection ☐ Inheritance ☐ Artificial selection

[1]

1.2 What is selective breeding?

....

....

[1]

1.3 Suggest why dairy farmers might use selective breeding.

....

[1]

[Total 3 marks]

2 A dog breeder used selective breeding to produce a litter of puppies that all had a good, gentle temperament. (Grade 6-7)

2.1* Describe how the breeder could have achieved this, starting from a mixed population of dogs.

....

....

....

....

....

....

[4]

2.2 Suggest why the puppies may be more susceptible to genetic defects.

....

....

....

[2]

2.3 Suggest why a new disease might be an issue for the puppies produced using selective breeding.

....

....

....

[3]

[Total 9 marks]

Genetic Engineering

Warm-Up

Draw circles to show whether the statements below are **true** or **false**.

Genetic engineering can only be carried out on plants.	True / False
Genetic engineering has been proven to be 100% risk free.	True / False
Crops that have been genetically engineered are already being grown.	True / False
Vectors in genetic engineering can be bacterial plasmids.	True / False

1 Genetic engineering is being investigated for use in a wide variety of applications. Grade 6-7

1.1 What is genetic engineering?

....

....

[2]

The process of genetic engineering has several steps.

1.2 The useful gene is first isolated from an organism's DNA.
Explain how this is done.

....

....

[1]

1.3 The gene is then inserted into the target organism's genome.
Explain how this is achieved so that the organism develops with the desired characteristics.

....

....

....

....

[3]

1.4 Scientists are currently investigating the applications of genetic engineering in medicine.
Give **two** examples of how genetic engineering has been used to treat human diseases, or how it could potentially be used.

....

....

....

....

[2]

[Total 8 marks]

2 Genetic engineering can be used to alter the genes and characteristics of food crops. The resulting crops are known as GM crops.

Grade 6-7

2.1 What does 'GM' stand for when referring to crops that have been genetically engineered? Tick **one** box.

☐ genetically manufactured ☐ genetically mutated ☐ genetically modelled ☐ genetically modified

[1]

2.2 GM crops are often altered to increase their yield. One way in which this can be achieved in some crops is by modifying their genes to make them produce larger fruit. Suggest **two** other ways in which a crop plant's genes can be altered to increase its yield.

....

....

[2]

A scientist was researching the effect of a genetic modification on fruit size in a species of plant. He first grew a normal individual of the species in controlled conditions (Plant 1). He then measured the circumferences of the 10 largest fruits after a set amount of time. He repeated these steps with a genetically modified individual from the same species (Plant 2). **Table 1** shows the results.

Table 1

	Fruit Circumference (cm)									
Plant 1	16.4	16.8	15.9	16.2	15.7	16.4	16.3	16.0	15.9	16.0
Plant 2	20.2	20.4	19.8	19.6	20.4	20.6	20.2	19.9	20.1	20.0

2.3 Use the data in **Table 1** to calculate the mean fruit circumference for each plant.

Plant 1 = ... cm Plant 2 = ... cm

[2]

2.4 Calculate the percentage change in mean fruit circumference between Plant 1 and Plant 2.

.............%

[2]

2.5 Not everyone thinks that GM crops are a good idea. Give **one** concern that people may have about GM crops.

....

....

[1]

[Total 8 marks]

Cloning

1 Plants can be cloned by tissue culture or by taking cuttings. Grade 4-6

1.1 Which of the following statements is correct?
Tick **one** box.

- [ ] Taking cuttings is an older and simpler method than tissue culture.
- [ ] Tissue culture is a newer and simpler method than taking cuttings.
- [ ] Taking cuttings is a newer and more complicated method than tissue culture.
- [ ] Tissue culture is an older and more complicated method than taking cuttings.

[1]

Figure 1 shows the process of tissue culture.

Figure 1

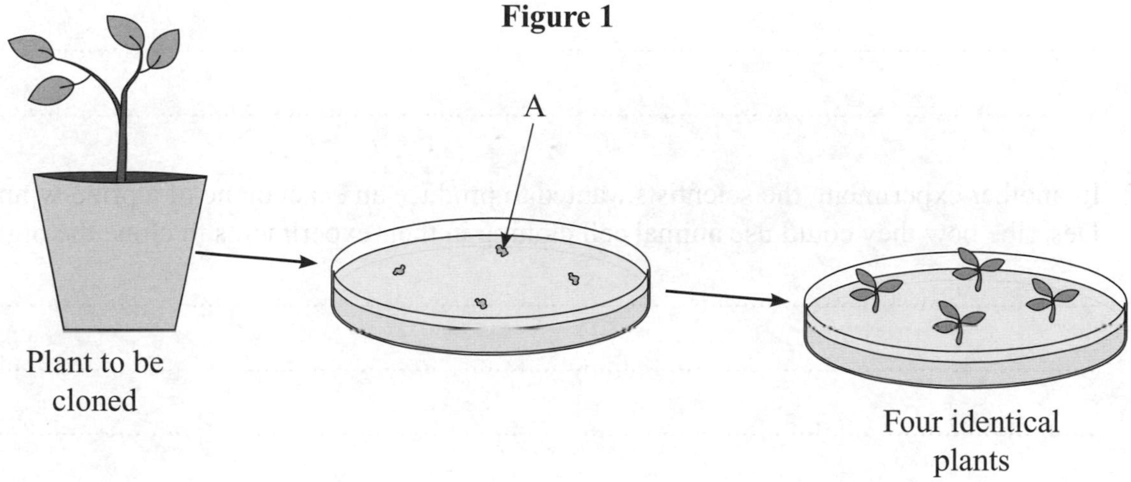

A

Plant to be cloned

Four identical plants

1.2 What is in the Petri dish in **Figure 1**, labelled A?

....

[1]

[Total 2 marks]

2 Plant nurseries sometimes use tissue culture to produce lots of identical plants to sell. Grade 6-7

2.1 Give **one** other use of plant tissue culture.

....

[1]

2.2 Explain a potential risk to plant nurseries of using cloning methods to produce stock.

....

....

....

....

[3]

[Total 4 marks]

3 Scientists were carrying out some cloning experiments using pigs. Grade 7-9

3.1 In one experiment, they wanted to produce a litter of genetically identical piglets using parents with desirable characteristics.
Describe how they could use embryo transplants for this purpose.

....

....

....

....

....

....

....

....

....

[5]

3.2 In another experiment, the scientists wanted to produce an exact clone of a prize-winning pig.
Describe how they could use animal cell cloning in their experiments to clone the pig.

....

....

....

....

....

....

....

....

....

....

....

....

....

[5]

[Total 10 marks]

Exam Practice Tip

Make sure you've got the two different types of animal cloning clear in your head. With all these egg cells and embryos flying around, it's dead easy to get the steps involved in each method jumbled up. Try writing all of the steps for each method down, checking them and then writing them out again. Do this a few times until you can remember every step.

Fossils

Warm-Up

Draw circles to show whether the statements below are **true** or **false**.

Fossils are all between 100 and 1000 years old.	True / False
Fossils are the remains of organisms.	True / False
Fossils are often found in rocks.	True / False

1 The fossil record provides an account of how much different organisms have changed over time. Fossils can be formed in three ways.

1.1 **Figure 1** shows a fossilised insect preserved in amber. Amber is fossilised tree sap. The insect became trapped in the sap as it fed from the tree.

Figure 1

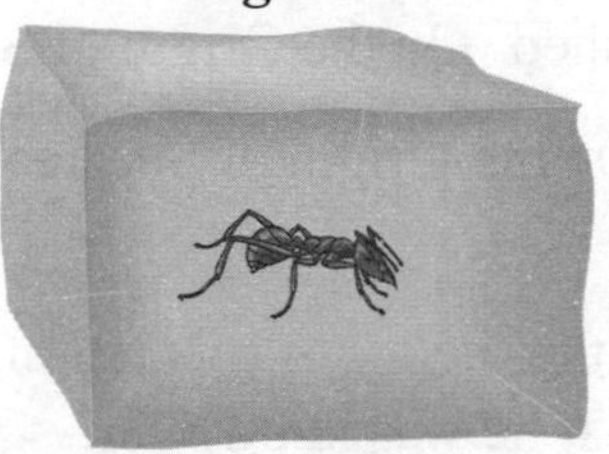

Explain how insects trapped in amber become fossilised rather than decaying.

....

....

[2]

1.2 Describe **two** other ways that fossils are formed.

....

....

....

....

[2]

1.3 Scientists are unable to use the fossil record as conclusive evidence to support or disprove theories on how life on Earth first began. Explain why this is the case.

....

....

....

....

....

....

[3]

[Total 7 marks]

Speciation

1 Alfred Russel Wallace and Charles Darwin were British biologists who carried out work on how species form and evolve. Grade 4-6

1.1 Alfred Russel Wallace and Charles Darwin proposed that evolution took place by one specific process. What is the name of that process?
Tick **one** box.

☐ Normal selection ☐ Natural Variation ☐ Natural selection ☐ Normal variation

[1]

1.2 Which of the following statements is correct?
Tick **one** box.

☐ Alfred Russel Wallace published 'On the Origin of Species' in 1859.
☐ Charles Darwin published 'On the Origin of Species' in 1859.
☐ Alfred Russel Wallace published 'On the Origin of Species' in 1851.
☐ Charles Darwin published 'On the Origin of Species' in 1851.

[1]

1.3 Alfred Russel Wallace worked on the theory of speciation.
State **one** other area of research that he was known for.

....

[1]

[Total 3 marks]

2 A population of a species is split up after its habitat is flooded.
Two isolated populations of the species form.
Explain how this could lead to the development of a new species.
Refer to variation and natural selection in your answer. Grade 7-9

....

....

....

....

....

....

....

....

....

....

....

[Total 5 marks]

Antibiotic-Resistant Bacteria

1 Antibiotic resistance in bacteria is becoming an increasing problem in medicine. This is partly due to the overuse of antibiotics. (Grade 4-6)

1.1 The overuse of antibiotics is sometimes caused by them being prescribed inappropriately. Give **two** examples of antibiotics being prescribed inappropriately.

....

....

[2]

1.2 Explain why patients prescribed a course of antibiotics should always complete the full course.

....

....

....

[3]

[Total 5 marks]

2 Antibiotic-resistant strains of bacteria are harder to treat because the conventional antibiotics used to kill them are now ineffective. New antibiotics are being developed, but it's unlikely that we'll be able to keep up with the emergence of new resistant strains. (Grade 6-7)

2.1 Explain why the development of antibiotics is unlikely to keep up with the emergence of new antibiotic-resistant bacteria.

....

....

[2]

2.2 Explain how antibiotic-resistant strains of bacteria develop and spread.

....

....

....

....

....

....

....

....

....

[5]

[Total 7 marks]

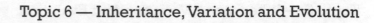

Classification

1 Evolutionary trees show how scientists think that organisms are related to each other. **Figure 1** shows the evolutionary tree for species A – K.

Figure 1

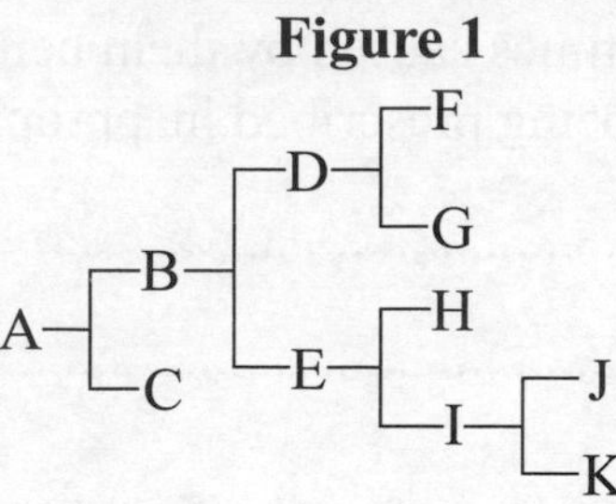

1.1 Give **two** pieces of information that scientists use to prepare evolutionary trees for living and extinct organisms.

....

[2]

1.2 Which species is the most recent common ancestor of species G and species J?

....

[1]

1.3 Which pair of species, G and H or J and K are more distantly related?

....

[1]

[Total 4 marks]

2 Organisms used to be classified into groups using the Linnaean system.

2.1 What is the correct order for the groups of the Linnaean system, from largest to smallest? Tick **one** box.

☐ kingdom, phylum, class, order, family, genus, species

☐ species, genus, class, phylum, order, family, kingdom

☐ kingdom, family, phylum, order, class, genus, species

☐ species, class, genus, family, order, phylum, kingdom

[1]

A new classification system, known as the three-domain system, was proposed in the 1990s. In this system, organisms are first divided into domains.

2.2 What is the name of the scientist who proposed the three-domain system?

....

[1]

2.3 Which of the domains includes primitive bacteria often found in extreme environments?

....

[1]

2.4 Other than fungi, state **three** types of organisms found in the domain Eukaryota.

....

[3]

[Total 6 marks]

Competition

1 The plants in a community are often in competition with each other for water and mineral ions. Grade 4-6

1.1 Where do plants obtain water and mineral ions from?

....

[1]

1.2 Name **two** other factors that plants often compete with each other for.

....

....

[2]

1.3 The animals in a community also compete with each other.
State **three** factors animals compete with each other for.

....

....

....

[3]

[Total 6 marks]

2 Within a community, each species depends on other species for things such as food, shelter, pollination and seed dispersal. Grade 6-7

2.1 What is this type of relationship called?

....

[1]

Blue tits are relatively common birds that live in woodland communities.
Blue tits feed on caterpillars. Caterpillars live and feed on plants.

2.2 If the caterpillars were removed from the community, suggest what might happen to the numbers of blue tits and plants. Explain your answers.

....

....

....

....

[4]

2.3 Some communities are not stable.
Explain fully what is meant by the term 'stable community'.

....

....

[2]

[Total 7 marks]

Abiotic and Biotic Factors

1 In an ecosystem, there will be both biotic and abiotic factors. Grade 4-6

1.1 Which of the following statements is correct? Tick **one** box.

- [ ] Light intensity, temperature and carbon dioxide level are all examples of biotic factors.
- [ ] Availability of food, carbon dioxide level and pathogens are all examples of abiotic factors.
- [ ] Light intensity, temperature and carbon dioxide level are all examples of abiotic factors.
- [ ] Availability of food, light intensity and pathogens are all examples of biotic factors.

[1]

1.2 Suggest **one** abiotic factor that could affect the distribution of animals living in aquatic areas.

....

[1]

1.3 Suggest **two** abiotic factors that could affect the distribution of plants growing in soil.

....

....

[2]

[Total 4 marks]

2 Red squirrels are native to southern Britain. When grey squirrels were introduced into the same area, the number of red squirrels declined. Grade 6-7

Suggest why the number of red squirrels declined.

....

....

....

[Total 3 marks]

3 Grasses make their own food by photosynthesis. In grassland communities, the grass leaves provide insects with shelter, a place to breed and a source of food. Visiting birds feed on insects. Grade 6-7

The birds that visit the grassland to feed become infected with a new pathogen that eventually kills them. What would you expect to happen to the number of grass plants? Explain your answer.

....

....

....

....

[Total 3 marks]

Adaptations

1 Some organisms live in environments that are very extreme, such as environments with a high salt concentration. Grade 4-6

1.1 What name is given to organisms that live in extreme environments?

....

[1]

1.2 Name **one** group of organisms that can live in deep sea vents where temperatures are very high.

....

[1]

1.3 Describe **one** extreme condition, other than a high salt concentration or a high temperature, that some organisms can tolerate.

....

[1]

[Total 3 marks]

2 Organisms are adapted to the conditions in which they live. **Figure 1** shows a camel and **Figure 2** shows a cactus. Both camels and cacti live in hot, dry desert conditions. Grade 6-7

Figure 1

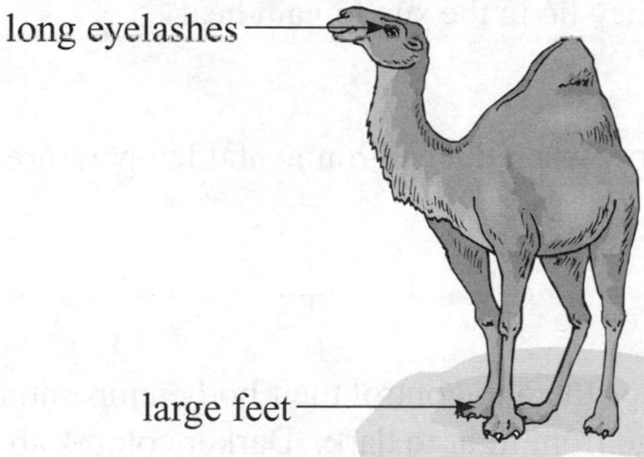

Figure 2

2.1 Suggest how each of the features in **Figure 1** allow the camel to live in desert conditions.

Long eyelashes ....

....

Large feet ....

....

[2]

2.2 Suggest how having spines instead of leaves allows cacti to live in desert conditions.

....

[1]

2.3 Suggest how having a swollen storage stem allows cacti to live in desert conditions.

....

[1]

Cacti can have two types of roots — shallow, wide-spreading roots or long, deep roots.

2.4 For each of these types of root, suggest how the cacti are better adapted to live in desert conditions.

....

....

[2]

[Total 6 marks]

3 Lizards gain most of their heat from the environment. This means that their body temperatures change with the temperature of the environment. **Figure 3** shows a lizard.

Figure 3

Lizards control their body temperatures by adapting their behaviour to changes in the environment. For example, in the early morning they lie in the sun to gain heat and only then can they become active.

3.1 Suggest what behavioural adaptation lizards might show when the environmental temperature becomes too hot.

....

[1]

Some lizards also have a structural adaptation that helps them to control their body temperature. They can change the colour of their skin within a range from light to dark. Darker colours absorb more heat than lighter colours.

3.2 Would you expect a lizard in cold conditions to have a dark or light coloured skin?

....

[1]

3.3 There are different types of adaptation, including behavioural and structural. Name **one** other type of adaptation that organisms can have.

....

[1]

[Total 3 marks]

Exam Practice Tip

All of the organisms on these pages are different and have different adaptations to their environment. You could be given information on any organism in the exam but, remember, each feature that the organism has gives it an advantage for living in its environment. You just need to have a think about what that advantage could be. Simple.

Food Chains

Warm-Up

On the food chain below, label the producer and the secondary consumer.

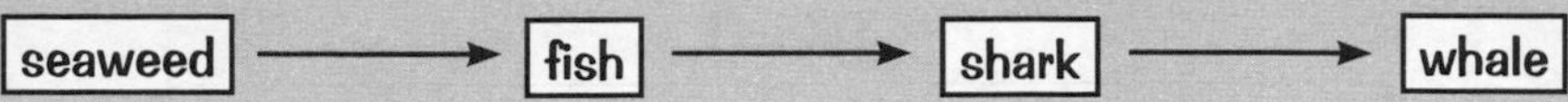

1 **Figure 1** shows an example of a woodland food chain.

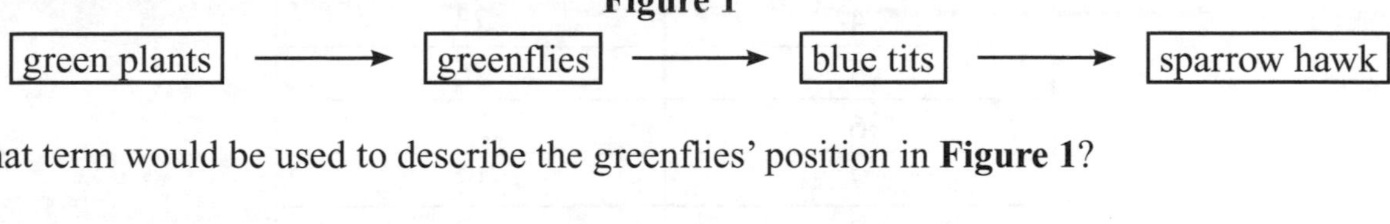

Figure 1

1.1 What term would be used to describe the greenflies' position in **Figure 1**?

....

[1]

1.2 Green plants produce biomass for the rest of the food chain. Explain how they do this.

....

....

[3]

[Total 4 marks]

2 In a stable community, the numbers of predators and prey rise and fall in cycles as shown in **Figure 2**. In this cycle the predator is a lynx and the prey is a snowshoe hare.

Figure 2

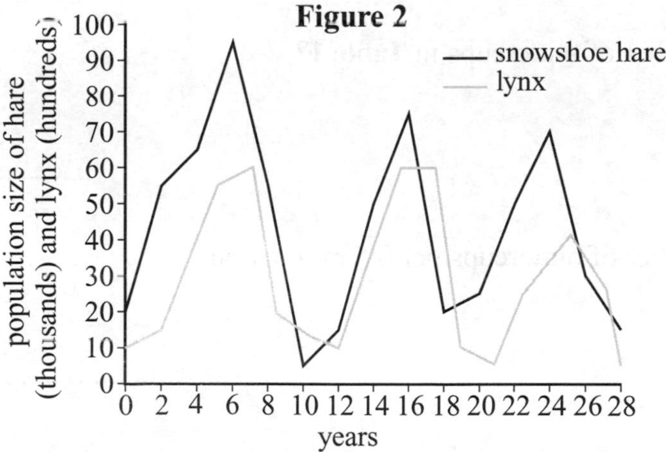

2.1 Describe and explain what happens to the number of lynx between years 4 and 6.

....

....

[2]

2.2 What causes the number of snowshoe hares to fall between years 6 and 10?

....

[1]

[Total 3 marks]

PRACTICAL

Using Quadrats

1 A group of students used a 0.5 m^2 quadrat to investigate the number of buttercups growing in a field. They counted the number of buttercups in the quadrat in ten randomly selected places. **Table 1** shows their results.

Table 1

Quadrat Number	Number of buttercups
1	15
2	13
3	16
4	23
5	26
6	23
7	13
8	12
9	16
10	13

1.1 Why is it important that the quadrats were placed randomly in the field?

.... *[1]*

1.2 What is the modal number of buttercups in **Table 1**?

.... buttercups *[1]*

1.3 What is the median number of buttercups in **Table 1**?

.... buttercups *[1]*

1.4 Calculate the mean number of buttercups per 0.5 m^2 quadrat.

.... buttercups per 0.5 m^2 *[1]*

1.5 The total area of the field was 1750 m^2.
Estimate the number of buttercups in the whole of the field.

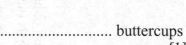

.... buttercups *[3]*

[Total 7 marks]

Using Transects

1 A transect was carried out from the edge of a small pond, across a grassy field and into a woodland. The distributions of four species of plant were recorded along the transect, along with the soil moisture and light levels. **Figure 1** shows the results.

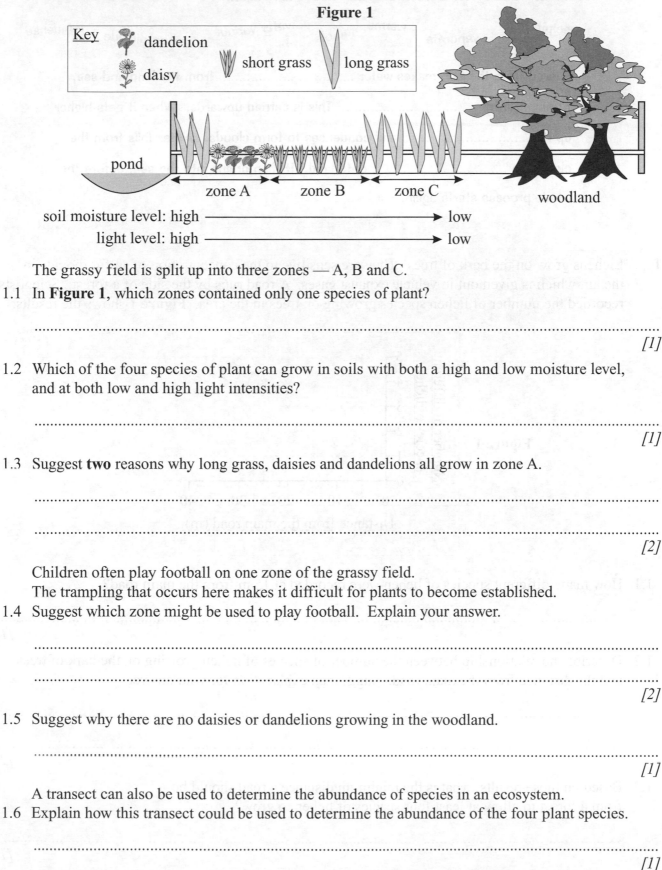

Figure 1

Key
dandelion
daisy
short grass
long grass

pond
zone A
zone B
zone C
woodland

soil moisture level: high → low
light level: high → low

The grassy field is split up into three zones — A, B and C.

1.1 In **Figure 1**, which zones contained only one species of plant?

....

[1]

1.2 Which of the four species of plant can grow in soils with both a high and low moisture level, and at both low and high light intensities?

....

[1]

1.3 Suggest **two** reasons why long grass, daisies and dandelions all grow in zone A.

....

....

[2]

Children often play football on one zone of the grassy field.
The trampling that occurs here makes it difficult for plants to become established.

1.4 Suggest which zone might be used to play football. Explain your answer.

....

....

[2]

1.5 Suggest why there are no daisies or dandelions growing in the woodland.

....

[1]

A transect can also be used to determine the abundance of species in an ecosystem.

1.6 Explain how this transect could be used to determine the abundance of the four plant species.

....

[1]

[Total 8 marks]

Environmental Change & The Water Cycle

Warm-Up

Choose from the words below to complete the sentences about the water cycle. Some words may not be used at all.

precipitation evaporate warms cools water vapour carbon dioxide condense

Energy from the Sun makes water .... from the land and sea, turning it into .... This is carried upwards. When it gets higher up it .... and condenses to form clouds. Water falls from the clouds as .... onto land. It then drains into the sea, before the whole process starts again.

1 Lichens grow on the bark of trees. They are sensitive to the concentration of sulfur dioxide in the air, which is given out in vehicle exhaust gases. A road runs by the side of a forest. Scientists recorded the number of lichen species growing on trees in the area. **Figure 1** shows the results.

Figure 1

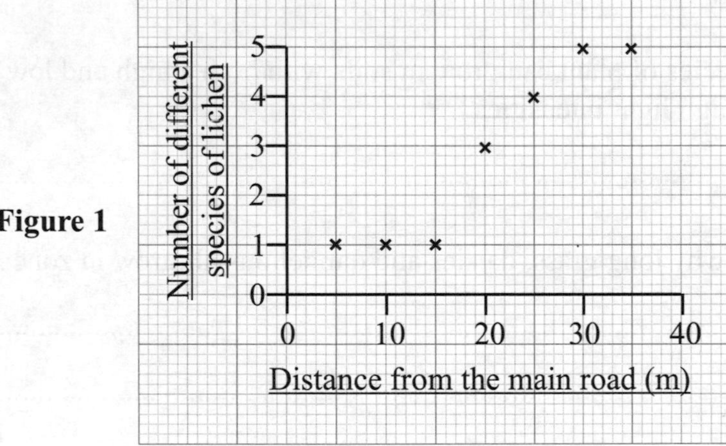

1.1 How many different species of lichen were recorded at 15 m from the main road?

....

[1]

1.2 Describe the relationship between the number of species of lichen growing on the bark of trees and the distance from the main road. Suggest an explanation for your answer.

....

....

[2]

1.3 Based on these results, what is the minimum distance a road should be from a forest to allow at least four species of lichen to grow?

....

[1]

[Total 4 marks]

The Carbon Cycle

1 **Figure 1** shows an unfinished diagram of the carbon cycle. 

Figure 1

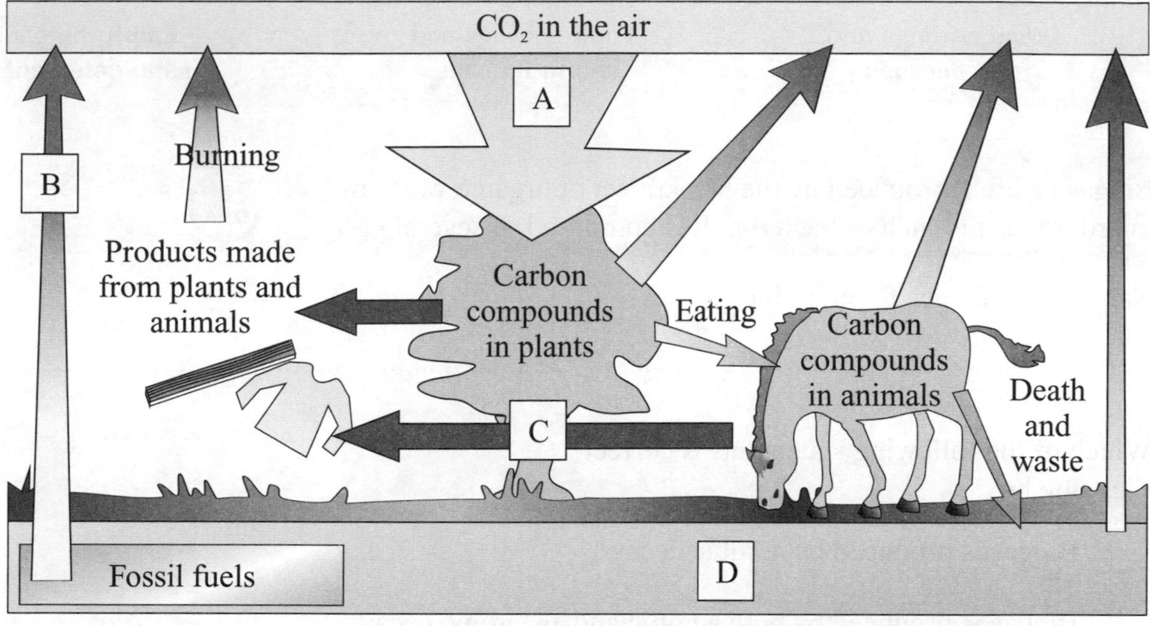

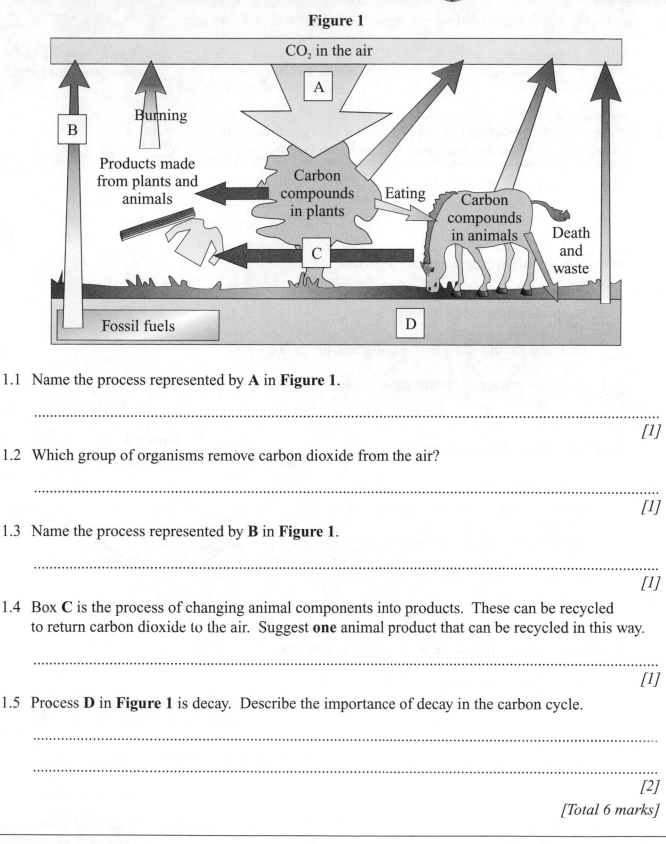

1.1 Name the process represented by **A** in **Figure 1**.

....

[1]

1.2 Which group of organisms remove carbon dioxide from the air?

....

[1]

1.3 Name the process represented by **B** in **Figure 1**.

....

[1]

1.4 Box **C** is the process of changing animal components into products. These can be recycled to return carbon dioxide to the air. Suggest **one** animal product that can be recycled in this way.

....

[1]

1.5 Process **D** in **Figure 1** is decay. Describe the importance of decay in the carbon cycle.

....

....

[2]

[Total 6 marks]

Exam Practice Tip

In the exam you could be tested on any part of the carbon cycle, so make sure you know the whole of it and not just bits of it. Try sketching the whole cycle out and make sure you can link each bit together. Don't have your arrows going the wrong way round, and make sure you understand why the carbon is moving around, e.g. because of respiration. Sorted.

Decay

Warm-Up

Circle which household wastes are only made up of organic matter.

Grass cuttings and food peelings. | Food peelings and empty tin cans. | Empty tin cans and plastic detergent bottles.

1 Biogas is a fuel produced by the breakdown of organic waste by microorganisms such as bacteria. It is composed of several gases.

1.1 Name the main gas found in biogas.

.... [1]

1.2 Which of the following statements is correct?
Tick **one** box.

☐ Biogas is produced by aerobic decay.

☐ Biogas is produced by both aerobic and anaerobic decay.

☐ Biogas is produced by anaerobic decay.

[1]

Figure 1 shows a biogas generator.

Figure 1

organic waste → biogas
digested material

1.3 The biogas generator has been built underground. Suggest **one** reason why.

....

.... [1]

Gardeners use the breakdown of organic waste to produce compost.

1.4 What do gardeners use compost for?

.... [1]

1.5 Describe **three** conditions that gardeners should use to produce compost quickly.

....

....

.... [3]

[Total 7 marks]

Investigating Decay

PRACTICAL

1 A student was investigating the effect of temperature on the decay of milk by an enzyme.

1. He added phenolphthalein to a sample of alkaline milk in a test tube and placed it in a water bath.
2. When the mixture in the tube reached the desired temperature, he added some lipase enzyme solution.
3. As the lipase broke down the milk, the pH of the mixture dropped and a colour change occurred.
4. He timed how long it took for the colour change to occur at four different temperatures. **Table 1** shows his results.

Table 1

	Time taken for colour change to occur (s)			
Temperature (°C)	1st repeat	2nd repeat	3rd repeat	mean
10	292	299	291	294
20	256	257	261	258
30	240	235	239	238
40	217	224	219	

1.1 Phenolphthalein acts as an indicator dye. It is responsible for the visible colour change that occurred. Describe how the colour of the indicator would have changed as pH decreased.

....

[1]

1.2 Calculate the mean time taken for the colour change to occur at 40 °C.

Mean = ... seconds

[1]

Figure 1

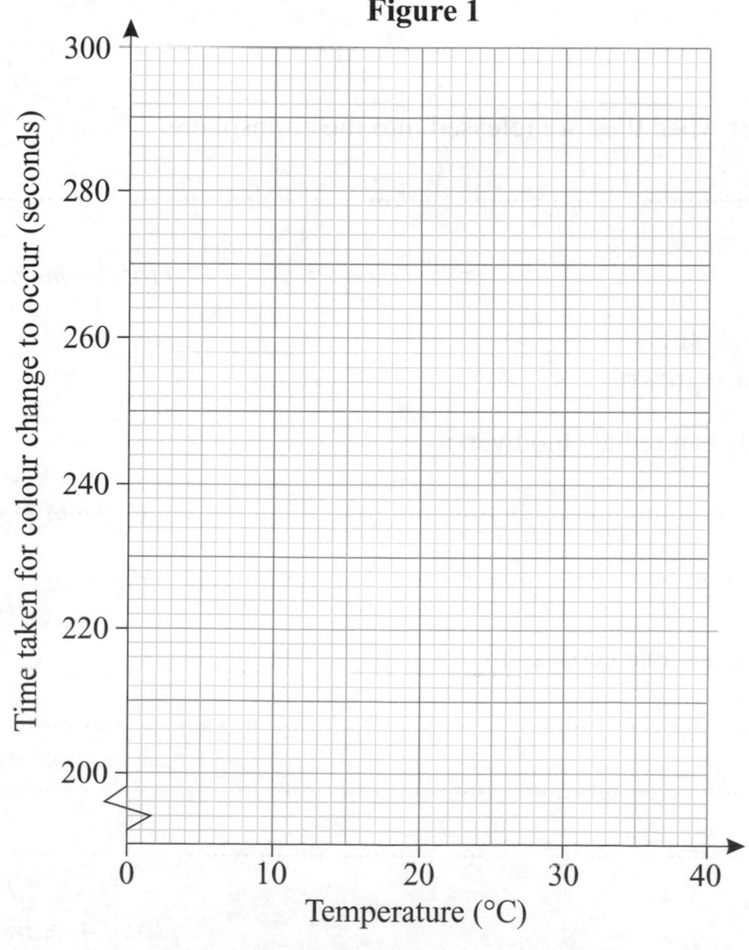

Figure 1 is an incomplete graph to show the mean time taken for the colour change to occur against temperature.

1.3 Plot the mean time taken for the colour change to occur at each temperature on **Figure 1**.

[2]

1.4 Complete **Figure 1** by drawing a curve of best fit.

[1]

1.5 Use your curve to predict how many seconds it would take for the colour change to occur at 35 °C.

....

[1]

[Total 6 marks]

Biodiversity and Waste Management

1 Many human activities have an impact on biodiversity. Grade 4-6

1.1 Define biodiversity.

....

....

[1]

1.2 Suggest **one** human activity that reduces biodiversity.

....

[1]

[Total 2 marks]

2 The global population is using an increasing amount of resources. Grade 4-6

2.1 State **two** reasons why humans are using more resources.

....

....

[2]

If waste is not handled correctly, pollution levels in water and in the air will increase.

2.2 State **two** ways that water can become polluted.

....

....

[2]

2.3 State **two** substances that pollute the air when they are released into the atmosphere.

....

[2]

[Total 6 marks]

3 Herbicides are used by farmers to control the growth of weeds on land where crops are grown. Grade 6-7

3.1 Give **two** reasons why using a herbicide can reduce biodiversity.

....

....

[2]

3.2 Explain why a high biodiversity creates a stable ecosystem.

....

....

....

[2]

[Total 4 marks]

Global Warming

Warm-Up

Fill in the gaps in the passage below using the words on the right.
Not all words need to be used, but each word can only be used once.

The temperature of the Earth is a balance between the energy it gets from .... and the energy it radiates back into .... The .... in the atmosphere act like an insulating layer. They radiate some of the energy back towards the Earth. This .... the temperature of the planet.

insulating
the Sun
reflect
increases
the moon
space
gases
decreases
volcanoes

1 Global warming is caused by the increasing levels of 'greenhouse gases' in the atmosphere. Grade 6-7

1.1 Which of the following pairs of gases are the main contributors to global warming? Tick **one** box.

☐ carbon dioxide and sulfur dioxide

☐ carbon dioxide and methane

☐ sulfur dioxide and nitrogen dioxide

☐ nitrogen dioxide and methane

[1]

There are a number of consequences of global warming, including land becoming flooded.

1.2 Suggest why global warming might cause land to become flooded.

....

....

[2]

1.3 Suggest **two** other consequences of global warming.

....

....

[2]

[Total 5 marks]

Exam Practice Tip

Greenhouse gases cause the 'greenhouse effect', but don't go getting this muddled up with global warming. Without these gases trapping energy in Earth's atmosphere, it'd be too cold for us to survive. But the increasing amounts of these gases are increasing the amount of energy being trapped, and that's why we're getting all, ahem, hot and bothered about them.

Deforestation and Land Use

1 State **two** ways in which humans reduce the amount of land available for other animals and plants. Grade 4-6

....

....

[Total 2 marks]

2 Peat bogs are sometimes destroyed so that the peat can be burnt as fuel. Grade 6-7

2.1 Give **one** other reason why peat bogs are destroyed by humans.

....

[1]

2.2 Explain what problem burning peat can cause.

....

....

[2]

2.3 Explain the effect that the destruction of peat bogs has on biodiversity.

....

....

[2]

[Total 5 marks]

3 Biofuel production has caused large-scale deforestation. Grade 6-7

3.1 Explain why large-scale deforestation has been required to produce biofuels.

....

[1]

3.2 State **two** other reasons for large-scale deforestation.

....

....

[2]

[Total 3 marks]

4 Suggest and explain **two** harmful effects on the environment caused by the destruction of large areas of trees. Grade 7-9

....

....

....

....

[Total 4 marks]

Maintaining Ecosystems and Biodiversity

1 In some areas, programmes have been put in place to reduce the negative effects of human activity on ecosystems and biodiversity. Grade 4-6

1.1 Which of the following would reduce carbon dioxide emissions into the atmosphere? Tick **one** box.

- ☐ Setting up more breeding programmes for endangered species.
- ☐ Cutting down large areas of trees for housing development.
- ☐ Increasing the number of power stations.
- ☐ Burning fewer fossil fuels.

[1]

1.2 The government encourages people to recycle as much of their waste as possible. Suggest how this could help to protect ecosystems.

....

....

[2]

[Total 3 marks]

2 Monoculture is a form of agriculture in which only one type of crop is grown in a field. Grade 6-7

2.1 Suggest what effect monoculture has on biodiversity. Explain your answer.

....

....

[2]

2.2 Explain why farmers who grow crops using monoculture may be advised to leave strips of grassland and plant hedgerows around the edges of their crops.

....

[1]

[Total 3 marks]

3 Suggest why some people might be opposed to programmes that maintain biodiversity. Grade 7-9

....

....

....

....

....

....

....

....

[Total 4 marks]

Trophic Levels

Warm-Up

The diagram below shows some of the feeding relationships in a rocky shore environment. Sort the organisms into the columns of the table. Each organism belongs in only one column.

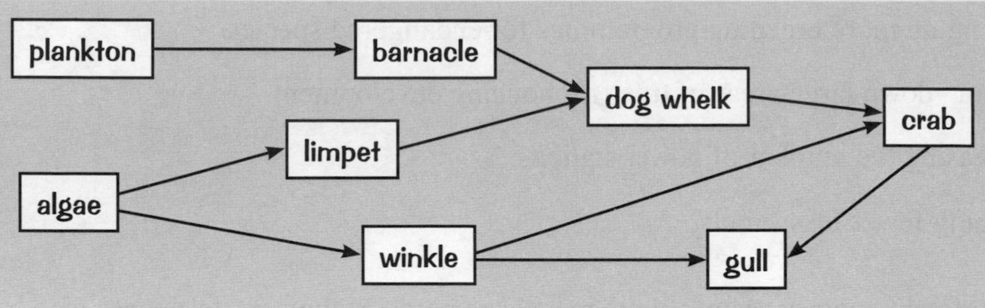

Producer	Herbivore	Carnivore

1 The trophic levels in a food chain can be represented by numbers, starting at level 1 with producers. (Grade 4-6)

1.1 Which of the following organisms would be found at level 2? Tick **one** box.

☐ photosynthetic organisms ☐ carnivores ☐ herbivores ☐ decomposers

[1]

1.2 Some carnivores eat other carnivores. Which level represents these carnivores? Tick **one** box.

☐ level 1 ☐ level 2 ☐ level 3 ☐ level 4

[1]

1.3 What is an apex predator?

....

[1]

[Total 3 marks]

2 Decomposers play an important role in ecosystems. (Grade 6-7)

Describe their role and explain how they carry it out.

....

....

....

....

[Total 4 marks]

Pyramids of Biomass

1 Pyramids of biomass can be constructed to represent the relative amount of biomass in each level of a food chain. 

Figure 1 shows a food chain from an area of oak woodland. The biomass values are given in arbitrary units.

Figure 1

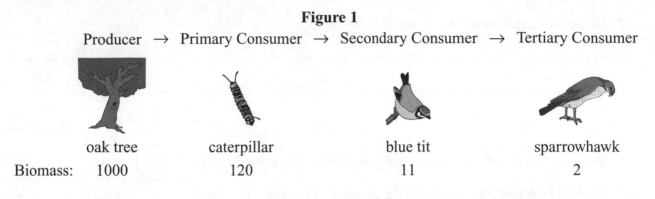

Producer → Primary Consumer → Secondary Consumer → Tertiary Consumer

	oak tree	caterpillar	blue tit	sparrowhawk
Biomass:	1000	120	11	2

1.1 Use the biomass values given in **Figure 1** to construct a pyramid of biomass.

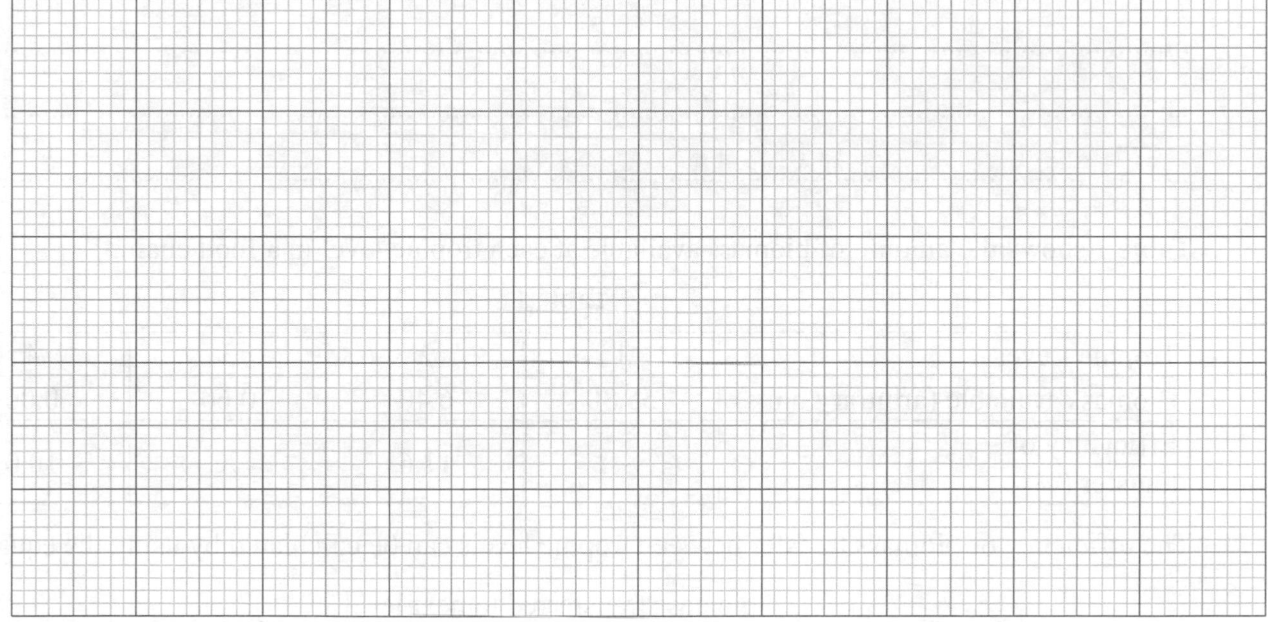

[4]

1.2 Use the biomass values from **Figure 1** to suggest why there are usually only four or five trophic levels in a food chain.

....

....

[2]

[Total 6 marks]

Exam Practice Tip

It's a good idea to always take a sharp pencil, ruler and eraser into the exam (as well as your lucky pen, of course). Other things that come in handy include: a calculator, a spare pen, a tissue, a bottle of water and a kitten (for the stress).

Biomass Transfer

1 Not all material that is eaten is used by the body. Grade 4-6

1.1 What happens to ingested material that does not get absorbed?

....

[1]

1.2 Name **two** substances lost as waste in urine.

....

....

[2]

[Total 3 marks]

2 There are losses of biomass at each trophic level in a food chain. Grade 6-7

2.1 Explain how respiration affects the amount of biomass that is transferred from one trophic level to the next.

....

....

....

....

[4]

Table 1 shows the amount of biomass available at each trophic level in a food chain.

Table 1

Trophic Level	1	2	3	4
Biomass available (arbitrary units)	55.30	6.40	0.60	0.06
Efficiency of transfer (%)	–	11.6	X	10.0

The efficiency of biomass transfer between trophic levels can be calculated by using the equation:

$$\text{efficiency} = \frac{\text{biomass transferred to next level}}{\text{biomass available at the previous level}} \times 100$$

2.2 Calculate the value of **X** in **Table 1**.

Efficiency of biomass transfer = ... %

[2]

2.3 Calculate the mean efficiency of biomass transfer between the trophic levels in **Table 1**.

Mean efficiency of biomass transfer = ... %

[2]

[Total 8 marks]

Food Security and Farming

1 Several factors affect food security. (Grade 4-6)

1.1 Which of the following factors is **not** a threat to food security?
Tick **one** box.

☐ A new disease that affects crops.

☐ Conflict over resources.

☐ Decreasing birth rate.

☐ High costs of farming.

[1]

1.2 Give **one** example of an environmental change that could affect food production.

....

[1]

[Total 2 marks]

2 Fish stocks around the world are monitored regularly. (Grade 6-7)

2.1 Explain why it is important to maintain fish stocks at a level where breeding continues.

....

[1]

2.2 What has been done to try to conserve fish stocks at a sustainable level?

....

....

....

[2]

[Total 3 marks]

3 Fish can be intensively reared in fish farms, as shown in **Figure 1**. (Grade 6-7)

Figure 1

3.1 Explain why the movement of fish reared intensively is restricted.

....

....

[2]

Species of fish that are intensively reared include salmon.
Salmon is a carnivorous fish that needs a high-protein diet.

3.2 Suggest why carnivorous species of fish are less efficient to farm than plant-eating fish.

....

....
[2]

3.3 Intensive rearing is also known as 'factory farming'. Animals are kept close together.
Suggest **one** disadvantage of factory farming techniques.

....
[1]

[Total 5 marks]

4 A scientist researched the amount of animal feed needed to produce 1 kg of three different types of meats on a farm. **Table 1** shows the results. Grade 7-9

Table 1

Animal	Chicken	Pigs	Cattle
Amount of feed needed to produce 1 kg meat (kg)	2.1	4.1	10.5

4.1 Which animal is the most efficient food source? Explain your answer.

....

....
[2]

4.2 Calculate the ratio of the amount of feed needed to produce 1 kg of meat from chicken to the amount needed to produce 1 kg of meat from cattle. Give the ratio in its simplest form.

Ratio = ... : ....
[1]

4.3 Animals such as cattle, which are farmed for meat, can be fed using crops.
The global production and consumption of meat is increasing.
Suggest what effect increasing meat consumption may have on global food security.
Explain your answer.

....

....

....

....

....
[3]

[Total 6 marks]

Exam Practice Tip

When you're faced with a wordy or difficult-looking question, try underlining the key bits to help you focus on what is actually being asked (e.g. you could underline the two values that you need to extract from the table in 4.2 above). By finding the key words, you'll be less likely to mis-read the question or start writing about something totally irrelevant.

Biotechnology

1 Some organisms can be genetically modified to produce desired substances. Grade 4-6

1.1 Give **one** use for genetically modified bacteria.

....

[1]

1.2 Suggest **two** advantages of genetically modified crops.

....

....

[2]

[Total 3 marks]

2 Mycoprotein is used to produce protein-rich foods suitable for vegetarians. Grade 6-7

2.1 What is the name of the fungus that is used to produce mycoprotein?
Tick **one** box.

☐ *Candida* ☐ *Fusarium* ☐ *Penicillium* ☐ *E. coli*

[1]

The fungus is grown on a culture medium in a fermenter. **Figure 1** shows this process.

Figure 1

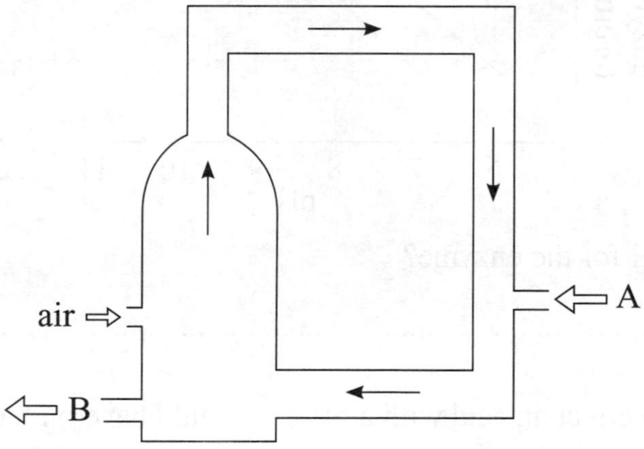

2.2 Which of the following options best describes the conditions in the fermenter?
Tick **one** box.

☐ anaerobic ☐ aerobic ☐ both anaerobic and aerobic

[1]

2.3 Substance **A** contains the sugar that the fungus feeds on. Name this sugar.

....

[1]

2.4 Mycoprotein is removed from the fermenter at point **B**.
Suggest the next stage required in the production process.

....

[1]

[Total 4 marks]

Mixed Questions

1 Alcohol is metabolised in the liver using alcohol dehydrogenase enzymes. Grade 4-6

1.1 State **one** function of the liver, other than alcohol metabolism.

....

[1]

1.2 Which of the following sentences about enzymes is **true**? Tick **one** box.

☐ Enzymes speed up chemical reactions in living organisms.

☐ Enzymes are used up in chemical reactions.

☐ Enzymes are products of digestion.

☐ Enzymes are the building blocks of all living organisms.

[1]

A scientist was investigating the effect of pH on the rate of activity of alcohol dehydrogenase. **Figure 1** shows a graph of his results.

Figure 1

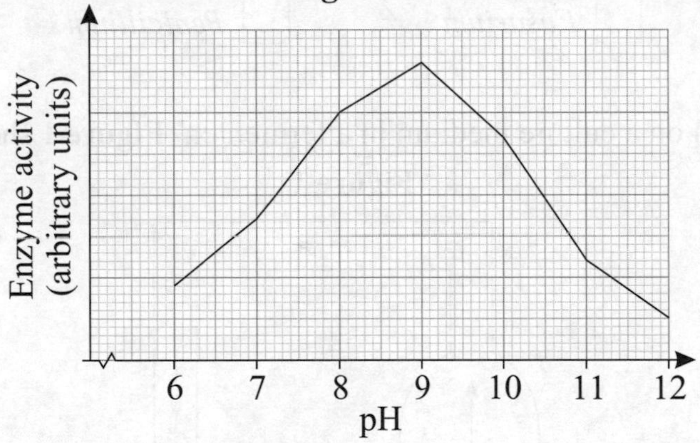

1.3 What is the optimum pH for the enzyme?

....

[1]

1.4 Suggest and explain the effect an acid with a pH of 1 would have on the enzyme.

....

....

....

[3]

1.5 Which of the following statements about alcohol is **not true**? Tick **one** box.

☐ Too much alcohol can cause liver disease.

☐ Alcohol is a risk factor for lung cancer.

☐ Alcohol can cause brain damage.

☐ Alcohol can affect unborn babies.

[1]

[Total 7 marks]

2 A group of students were investigating the effect of air flow on the rate of transpiration. They set up their apparatus as shown in **Figure 2**.

Figure 2

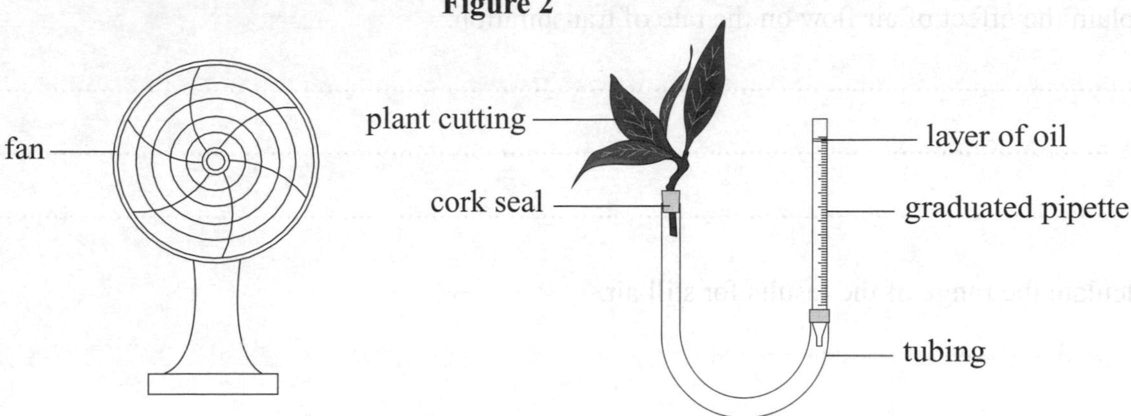

2.1 The tubing and graduated pipette were filled with water.
Suggest why a layer of oil was added to the surface of the water in the pipette.

....

[1]

The students recorded the change in the volume of water in the pipette over 30 minutes, in normal conditions. They repeated this five times. They then carried out these steps with the fan turned on to simulate windy conditions. **Table 1** shows their results.

Table 1

	Repeat	1	2	3	4	5	Mean
Water uptake in 30 minutes (cm^3)	Still Air	1.2	1.2	1.0	0.8	1.1	1.1
	Moving Air	2.0	1.8	2.3	1.9	1.7	1.9

2.2 Draw a bar chart to show the mean water uptake for still air and moving air.

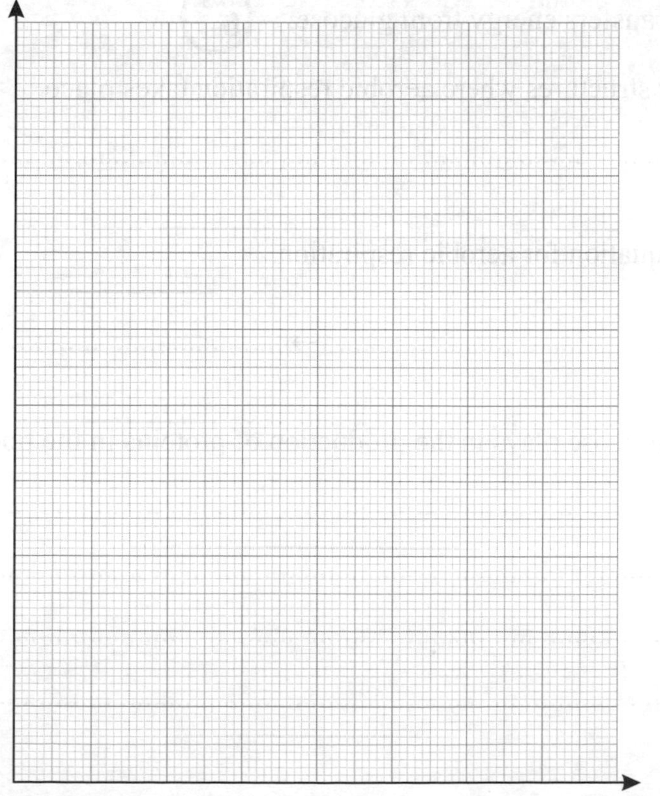

[2]

2.3 Describe the relationship between air flow around the plant and transpiration rate.

....

[1]

2.4 Explain the effect of air flow on the rate of transpiration.

....

....

....

[2]

2.5 Calculate the range of the results for still air.

Range = ... cm^3

[1]

2.6 The rate of transpiration can be calculated using the formula:

$$\text{rate of transpiration} = \frac{\text{mean volume of water uptake}}{\text{time taken}}$$

Calculate the rate of transpiration for the plant in moving air.
Give your answer in cm^3/hour.

.... cm^3/hour

[2]

[Total 9 marks]

3 Aerobic respiration transfers energy from glucose. Grade 6-7

3.1 Name the subcellular structures where aerobic respiration takes place.

....

[1]

3.2 Complete the word equation for aerobic respiration.

.... + ... → .... + ...

[2]

3.3 Outline the role that glucose plays in the production of proteins in the body.

....

....

....

....

[3]

[Total 6 marks]

4 The endocrine system uses hormones to produce effects within the body. Grade 6-7

4.1 Outline how a hormone travels from a gland to its target organ in the body.

....

....

[2]

The menstrual cycle is controlled by hormones.

Figure 3 shows the change in the levels of these hormones during one menstrual cycle. It also shows the change in the lining of the uterus.

Figure 3

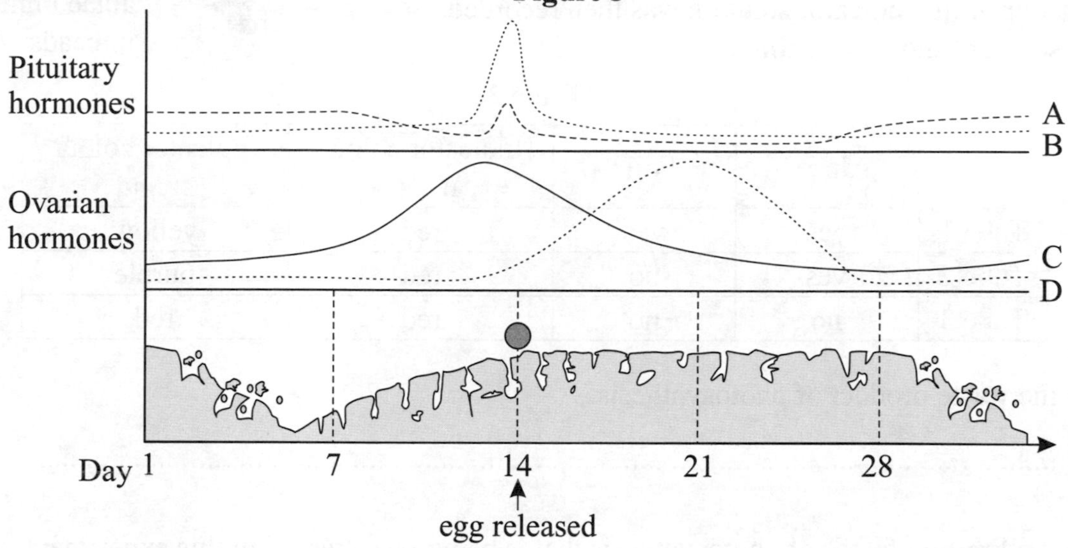

Pituitary hormones

Ovarian hormones

A

B

C

D

Day 1 7 14 21 28

egg released

4.2 Which line in **Figure 3** represents oestrogen? Tick **one** box.

☐ A ☐ B ☐ C ☐ D

[1]

4.3 Which line in **Figure 3** represents luteinising hormone? Tick **one** box.

☐ A ☐ B ☐ C ☐ D

[1]

4.4 What is the function of luteinising hormone?

....

[1]

4.5 Where in the body is progesterone produced?

....

[1]

4.6 Taking the combined pill keeps the level of oestrogen in the body constantly high. Explain how this reduces fertility.

....

....

[2]

[Total 8 marks]

5 A student was investigating the effect of limiting factors on the rate of photosynthesis by green algae.

PRACTICAL

The student set up two boiling tubes as shown in **Figure 4**.
She also set up a third tube that did not contain any algae.
The colour of the indicator solution changes as follows:

- At atmospheric CO_2 concentration, the indicator is red.
- At low CO_2 concentrations, the indicator is purple.
- At high CO_2 concentrations, the indicator is yellow.

Figure 4

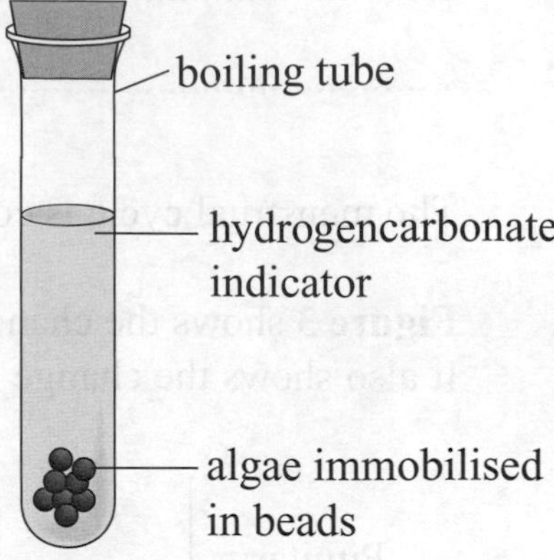

The student covered one of the boiling tubes containing algae with foil. All three tubes were left for several hours at room temperature with a constant light source.
The colour of the indicator solution was then recorded.
The results are shown in **Table 2**.

Table 2

	Algae?	Foil?	Indicator colour at start	Indicator colour at end
Tube 1	yes	yes	red	yellow
Tube 2	yes	no	red	purple
Tube 3	no	no	red	red

5.1 Name the waste product of photosynthesis.

....

[1]

5.2 Name the limiting factor of photosynthesis that is being investigated in this experiment.

....

[1]

5.3 At the end of the experiment, which tube has the highest carbon dioxide concentration?
Tick **one** box.

☐ Tube 1 ☐ Tube 2 ☐ Tube 3

[1]

5.4 Explain the results of Tube 1 and Tube 2.

....

....

....

....

....

[4]

5.5 Give **two** variables that needed to be controlled in this experiment.

....

....

[2]

A scientist investigating the effect of limiting factors on photosynthesis sketched the graph shown in **Figure 5**.

Figure 5

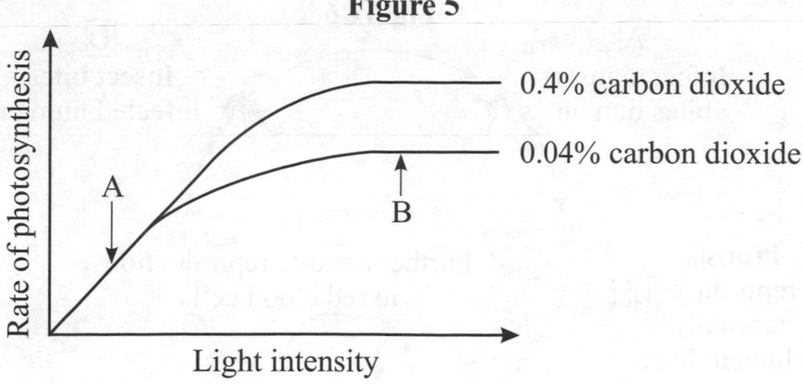

5.6 What is the limiting factor at point **A**? Explain your answer.

....

....
[2]

5.7 Name the limiting factor at point **B**.

....
[1]

[Total 12 marks]

6 In pea plants, seed shape is controlled by a single gene. Grade 7-9

The allele for round seed shape is R and the allele for wrinkled seed shape is r.
R is a dominant allele and r is recessive.

6.1 What is the genotype of a pea plant that is homozygous dominant for seed shape?

....
[1]

6.2 What is the phenotype of a pea plant that is heterozygous for seed shape?

....
[1]

6.3 Two pea plants were crossed. All of the offspring produced had the genotype **Rr**.
Construct a Punnett square to find the genotypes of the parent plants.

Genotypes: .... and ....
[3]

[Total 5 marks]

7 The life cycle of the protist that causes malaria is shown in **Figure 6**.

Figure 6

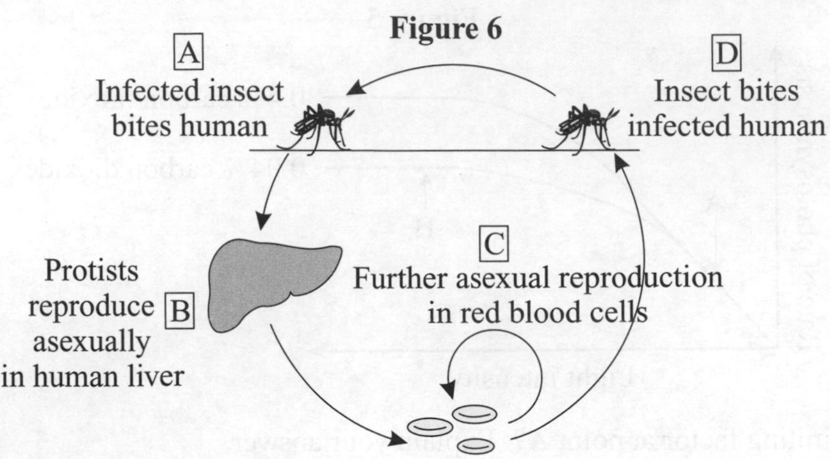

7.1 Suggest **one** method of blocking the protist's life cycle at point **A**.

.... *[1]*

7.2 Name the type of division that is occurring at point **B**.

.... *[1]*

7.3 Symptoms of malaria include feeling tired and lacking energy. The protist reproduction at point **C** destroys the red blood cells. Explain how this could cause these symptoms of malaria.

....

.... *[2]*

Malaria can be detected in a blood sample using a diagnostic stick, which works in a similar way to a pregnancy test. The stick is made from a strip of paper inside a plastic case. At one end of the stick, the paper contains antibodies (labelled with dye) that are specific to a malaria antigen — this is where a drop of blood and some colourless flushing agent are added. A positive result is revealed if a coloured line appears at a point further along the stick, as shown in **Figure 7**.

Figure 7

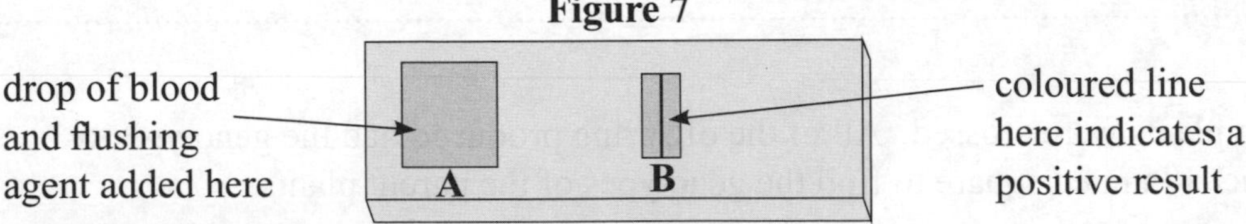

7.4 Suggest why some flushing agent is added with the blood at point **A** on the diagnostic stick.

.... *[1]*

7.5 The sample then moves along the stick. Suggest why a coloured line appears at point **B**.

....

....

....

.... *[4]*

[Total 9 marks]

BAQ41